# Learning from the COVID-19 Pandemic

COVID-19 is a highly contagious viral illness caused by severe acute respiratory syndrome SARS-CoV-2. It has had a devastating effect on the world's demographics with high morbidity and mortality worldwide. After the influenza pandemic of 1918, it has emerged as the most consequential global health crisis. After the first cases of this predominantly respiratory viral illness were first reported in Wuhan, Hubei Province, China, in late December 2019, SARS-CoV-2 rapidly disseminated across the world in a short span of time, compelling the World Health Organization (WHO) to declare it a global pandemic on March 11, 2020. The outbreak of COVID-19 has proven to be a world-wide unprecedented disaster. It has physically, psychologically, socially, and economically afflicted billions of people across the globe. Its transmission is significantly high. Serious postrecovery has been noticed in a large number of people. The virus is highly mutable and new and new strains are appearing, and many of them such as delta, BA1 and BA2 subvariants as well as their hybrids have been considered by the WHO as concerning. The virus has exhibited deleterious impacts on bodily systems other than the respiratory system (primary target) such as the brain, hematological system, liver, kidneys, endocrine system, etc. Right after its declaration as a pandemic by the WHO in March 2020, governments in various countries declared lockdowns to combat the spread of disease, causing major disruption to the lives of billions of people.

Besides the impact on health and healthcare systems, education was changed with the introduction of online and or hybrid systems to help students continue to learn. Though the pandemic has subsided now, the emergence of new variants continues and lifestyle changes such as online learning and work from home have continued.

Researchers who successfully mitigated the negative impact of social media and effectively used it for acceptance of medicinal or non-medicinal measures during pandemics by developing a real-time information sharing system and assembling a multidisciplinary team of experts to collect and analyze data from a variety of social media platforms across the global diaspora to better understand people's perceptions and attitudes, as well as to spot early warning signs of error and correct them before they proliferate. They also emphasized the necessity of addressing people's perceptions in order to increase awareness and education, so that social media may be used to promote public trust collaboration, and improved adherence to epidemic control measures. In totality the pandemic affected the environment and ecosystem as a whole positively due to a decrease in vehicles on roads and less movement of persons from one place to another. However, medical waste was increased and new measures were needed to handle it. People have had to change their habits in everyday life in order to live with the pandemic and protect themselves and others.

This volume focuses on the implications of COVID-19 on education, environment, and lifestyle. It includes chapters on the transformation of education systems and introduction of hybrid modes of education, impact on environment, management of solid wastes, and development of innovative gadgets and architectural designs to help deal with the pandemic. Other chapters cover diet, family systems, and adoption of new norms in pandemic times.

This book will be a valued resource for students, teachers, and researchers of social science and science as well as public health workers.

# Learning from the COVID-19 Pandemic

## Implications on Education, Environment, and Lifestyle

Edited By
R.C. Sobti, Vipin Sobti, and Aditi Sobti

CRC Press
Taylor & Francis Group
Boca Raton London New York

CRC Press is an imprint of the
Taylor & Francis Group, an **informa** business

First edition published 2023
by CRC Press
6000 Broken Sound Parkway NW, Suite 300, Boca Raton, FL 33487-2742

and by CRC Press
4 Park Square, Milton Park, Abingdon, Oxon, OX14 4RN

*CRC Press is an imprint of Taylor & Francis Group, LLC*

© 2023 selection and editorial matter, R.C. Sobti, Vipin Sobti, and Aditi Sobti; individual chapters, the contributors

ISBN: 978-1-032-41605-2 (hbk)
ISBN: 978-1-032-41606-9 (pbk)
ISBN: 978-1-003-35891-6 (ebk)

DOI: 10.1201/9781003358916

Typeset in Times New Roman
by Newgen Publishing UK

# Contents

# Preface

Coronavirus disease 2019 (COVID-19) is a highly contagious viral illness caused by severe acute respiratory syndrome SARS-CoV-2. It has had a devastating effect on the world's demographics with high morbidity and mortality worldwide and such impact still continues in various countries. It has emerged as the most consequential global health crisis since the era of the influenza pandemic of 1918. After the first cases of this predominantly respiratory viral illness were reported in Wuhan, Hubei Province, China, in late December 2019, SARS-CoV-2 rapidly disseminated across the world in a short span of time, compelling the World Health Organization (WHO) to declare it as a global pandemic on March 11, 2020. The outbreak of COVID-19 has proven to be a worldwide unprecedented disaster. Across the globe, billions of people have been physically, psychologically, socially, and economically affected by it. Its transmission is significantly high and have damaging worst post-recovery implications. The virus is highly mutable and new and new strains are appearing, and many of them have been declared as variants (such as delta, and now BA1 and BA2, their hybrids and others) of concern. The virus has exhibited deleterious impacts on systems other than the respiratory system (primary target organ) e.g., brain, liver, kidneys, endocrine system etc. Right after its declaration as pandemic by World Health Organization in March, 2020, governments in various countries declared lockdowns which no doubt was to combat the spread of disease, had all-round drastic effects.

Besides the impact on health and health care system, the education scenario was initially worst. The introduction of online and/or hybrid systems did help the students to complete their programs. Though the pandemic has subsided now, the emergence of new variants, the COVID-19 problem is not over yet. There had been an urgent need to maximize the use of online platforms so that students could not only finish their degrees, but also got prepared for the digitally oriented future environment. In such a pandemic condition, the notion of "work from home" had been more important in limiting the spread of COVID-19. During the pandemic, nations tried to devise innovative measures to guarantee that all children had continuous access to education. For efficient education delivery system, the policy makers tried to incorporate a varied range of persons from different backgrounds, including distant locations, marginalized and minority groups. Because online practice has proved to be quite beneficial to students, is being continued even after the lockdown.

Researchers who successfully mitigated the negative impact of social media and effectively used it for acceptance of medicinal or non-medicinal measures during pandemics by developing a real-time information sharing system and assembling a multidisciplinary team of experts to collect and analyze data from a variety of social media platforms across the global diaspora to better understand people's perceptions and attitudes, as well as to spot early warning signs of error and correct them before they proliferate. They also emphasized the necessity of addressing people's perceptions intended at increasing awareness and education, so that social media may be used to promote public trust collaboration, and improved adherence to epidemic control measures. In totality the pandemic affected the environment positively due to less transportation vehicles on roads and less movement of persons from one place to another. Though medical waste was increased and was difficult to handle, the frontline workers dedicatedly worked to take care of it. The pandemic completely changed our family life, life style and thought process. Now people have tried to evolve the new-normal in covid and post covid situations so as to live with it or with other pandemics. People have learnt to become more sensitive and responsible towards handling such situation.

The present work is second volume on the book *Learning from the Covid-19 Pandemic: Implications on Education, Environment and Lifestyle*. It concentrates on the implications on education, environment, and lifestyle and includes chapters on transformation of education systems including introduction of hybrid mode of education, impact on environment, management of solid wastes, and development of innovative gadgets to help the mankind and make architectural plans for better living in pandemic times. Chapters on transformation of life style including diet, family systems and adoption of new norms in pandemic times have also been included.

The students, teachers, researchers of social science and science as well as public health workers will find it as ready reckoner of information of their interest.

R. C. Sobti is thankful to the Indian National Science Academy for providing the platform as Senior Scientist to continue his academic pursuits.

Editors are thankful to Dr Aastha Sobti, Er Vineet, Er Ankit, and Miss Irene for their continuous support in compiling this book. Thanks to Dr Tejinder Kaur for her help in going through the text of a certain chapters.

**R. C. Sobti**
**Vipin Sobti**
**Aditi Sobti**

# Editor Biographies

**Ranbir Chander Sobti** is a Former Education Consultant Governor of Bihar, Senior Scientist (Indian National Science Academy), Former Vice Chancellor, Babasaheb Bhimrao Ambedkar University, Lucknow (UP), and Panjab University, Chandigarh. Professor Sobti is a Fellow of the Third World Academy of Sciences (TWAS), National Academy of Sciences, Indian National Science Academy, National Academies of Medical Sciences and Agricultural Sciences and of the Canadian Academy of Cardiovascular Diseases and is associated with many other Academic Associations and Institutions in the domain of higher education and research. The litany of honours showered on him includes, among others, the INSA Young Scientist Medal (1978), UGC Career Research Award, Punjab Rattan Award, JC Bose Oration Award and the Life Time Achievement Award of the Punjab Academy of Sciences, Zoological Society of India, and the Environment Academy of India, besides many other medals; awards of various reputed National and International Organizations. He was bestowed with Padmashri award by the Government of India in 2009.

**Vipin Sobti** is an educationist with more than 40 years of teaching experience in psychology and child development both at undergraduate and post-graduate levels. After completing her term as Principal, Government College Mohali, she was nominated as member of Education Tribunal, Punjab. She has published 4 books and a number of research papers. She has also guided a few students for the M.Sc. Degree in Child Development when she was teacher in the Home Science College, Chandigarh. She has attended and delivered talks in the conferences in India and abroad. She has travelled to UK, USA, Japan, Italy, Turkey, Singapore, Switzerland etc. and a number of other countries.

**Aditi Sobti** is a Senior Agile Project Manager is solution-oriented specialist with notable success in guiding cross functional teams, with a track record of meeting or exceeding corporate goals. She has demonstrated capacity to implement innovative web based and non-web-based applications in support of departmental goals. She has outstanding leadership abilities, decisive and detail oriented; able to coordinate and direct all phases of project-based efforts while managing, motivating, and leading teams. She can adopt and develop effective policies and procedures, project documentation and milestones, and technical/business specifications.

# Contributors

**Aditya Vikram Agarwal**
HS Srivastava Foundation for Science and
Society, Lucknow, India

**Anis Ahmad**
Department of Law, Babasaheb Bhimrao
Ambedkar University, Lucknow, Uttar Pradesh,
India

**Alka Agrawal**
Department of Information Technology,
Babasaheb Bhimrao Ambedkar University,
Lucknow, India

**Md Tarique Jamal Ansari**
Department of Information Technology,
Babasaheb Bhimrao Ambedkar University,
Lucknow, India

**Harish Kumar Banga**
Guru Nanak Dev Engineering College,
Ludhiana, Punjab, India

**Amita Chauhan**
UIET, Panjab University, Chandigarh, India

**Ramkesh Dalal**
Department of Zoology, Panjab University,
Chandigarh, India

**Parveen Goyal**
UIET, Panjab University, Chandigarh, India

**Akshi Goyal**
Department of Environment Studies,
Panjab University, Chandigarh, India

**Vibhash C. Jha**
Visva-Bharati University, Shantiniketan,
West Bengal, India

**Harmeet Kaur Kang**
Chitkara School of Health Sciences,
Chitkara University, Rajpura, Punjab, India

**Rupinder Bir Kaur**
University Business School, Panjab University,
Chandigarh, India

**Kanwal Roop Kaur**
University Business School, Panjab University,
Chandigarh, India

**Satvinderpal Kaur**
Department of Education, Panjab University,
Chandigarh, India

**Navneet Kaur**
Dev Samaj, College of Education, Chandigarh,
India

**Sakshi Kaushal**
UIET, Panjab University, Chandigarh, India

**Mohd. Rizwan Khan**
Department of English, Aligarh Muslim
University, Aligarh, Uttar Pradesh, India

**Raees Ahmad Khan**
Department of Information Technology,
Babasaheb Bhimrao Ambedkar University,
Lucknow, India

**Raj Kumar**
Banaras Hindu University, Varanasi,
Uttar Pradesh, India

**Santosh Kumar**
Department of Geology, Centre of Advanced
Study, Kumaun University, Nainital,
Uttarakhand, India

**Neelima R. Kumar**
Department of Zoology, Panjab University,
Chandigarh, India

**Harish Kumar**
UIET, Panjab University, Chandigarh, India

**Rajesh Kumar**
UIET, Panjab University, Chandigarh, India

**Pawan Kumar**
Department of Environmental Science,
Babasaheb Bhimrao Ambedkar University,
Raebareli Road, Lucknow, India

**Ashish Kumar Lamiyan**
Department of Zoology, Panjab University,
Chandigarh, India

**Jatinder Madan**
Chandigarh College of Engineering and
Technology (Degree Wing) Chandigarh, India

**Ravi K. Mahajan**
USOL, Panjab University, Chandigarh, India

**Kalpana K Mahajan**
Department of Statistics, Panjab University,
Chandigarh, India

**Nitish Mahajan**
UIET, Panjab University, Chandigarh, India

**Neeru Malik**
Dev Samaj, College of Education, Chandigarh,
India

**Rakesh Malik**
Dev Samaj, College of Education, Chandigarh,
India

**Jagdish C. Mehta**
Department of Sociology, D.A.V. College,
Chandigarh, India

**Suman Mor**
Department of Environment Studies,
Panjab University, Chandigarh, India

**Sahil Mor**
Department of Environmental Science and
Engineering, Guru Jambeshwar University of
School and Technology, Hisar, Haryana, India

**Mahendra Kumar Padhy**
Department of Mass Communication, Babasaheb
Bhimrao Ambedkar University, Lucknow, India

**Abhishek Pandey**
Department of Geology, Centre of Advanced
Study, Kumaun University, Nainital,
Uttarakhand, India

**Khaiwal Ravindra**
Department of Community Medicine and School
of Public Health, PGIMER, Chandigarh, India

**Aparna Sarin**
Uttarakhand State Council for Science &
Technology, Dehradun, Uttarakhand, India

**Tejinderpal Singh**
University Business School, Panjab University
Chandigarh, India

**Gurkirpal Singh**
IKG Punjab Technical University, Mohali
Campus, Panjab, India

**Rana Pratap Singh**
Department of Environmental Science,
Babasaheb Bhimrao Ambedkar University,
Raebareli Road, Lucknow, India

**Tanbir Singh**
Department of Environment Studies,
Panjab University, Chandigarh, India

**Komal Singh**
Babasaheb Bhimrao Ambedkar University,
Lucknow, Uttar Pradesh, India

**R. C. Sobti**
Department of Biotechnology,
Panjab University, Chandigarh, India

**Shilpi Verma**
Department of Library & Information Science,
Babasaheb Bhimrao Ambedkar University,
Lucknow, Uttar Pradesh, India

# 1 Teaching, Learning, and Caring in Post-COVID Era

*Shilpi Verma*

Department of Library & Information Science, Babasaheb Bhimrao Ambedkar University, VidyaVihar, Lucknow, Uttar Pradesh, India

## CONTENTS

## 1.1 INTRODUCTION

As the world grappled with the COVID-19 crisis and in this situation, children are also struggling with their ever-growing needs like food, health, medical care, physical safety, emotional support, mental wellbeing, teaching learning and many more things. It is well known that COVID-19 has affected to whole world but in comparison to urban areas, there is devasting effect on rural communities where many children are not getting enough food and unemployment is high. The digital divide is also high among rural populations and some places lack internet completely. Moreover, due to social distancing students are not able to board buses and other public transport, and as a result they can not reach schools and colleges. Teachers are trying hard to get connected to their students by utilizing various means, but still in India many parts of the country are struggling for internet access, electricity, etc. Now administrators are looking for people who can help them in filling these gaps between students and teachers. The administrator who are able to make revolutionary change, had implemented various methods so that the child may receive proper support and assistance in this pandemic situation.

DOI: 10.1201/9781003358916-1

1

## 1.2 HIGHER EDUCATION IN COVID ERA

The COVID pandemic forced universities across India and abroad to convert to online learning. Private universities in India had a quick transition, but the transition was very difficult for government institutions. Still, they are trying to adapt to the changes. There were long debates on the future of evaluation systems, the nature of classes and assessment, evaluation, and whether these should be conducted online. Faculty members tried to manage the various methods of online teaching, while students also looked for ways to keep up with their studies. Online education is not as easy as someone speaking into a microphone on one end and at the other someone listening through a laptop or mobile. Challenges are faced at both ends, as it is also not easy for teachers to speak in front of a camera without the physical presence of students and students lack the face-to-face time they are used to.

### 1.2.1 THE UNIVERSITIES

After encountering COVID-19 crises in India, various reputed universities like Delhi University (DU) and Jawaharlal Nehru University (JNU) shifted themselves from physical to online mode. Similarly, other private and government universities, colleges followed the same pattern for center the needs of the students and future of the nation.

### 1.2.2 THE IITs (INDIAN INSTITUTE OF TECHNOLOGY)

IITs also shifted to online classes and other forms of virtual learning like audio lectures. The teachers made themselves available online during the interactive sessions so that students could ask questions and get their questions answered. Teachers are augmenting Moodle, an open-source learning management system, with a variety of social media and online platforms to meet the needs of their students. The kind of courses being taught and the students' access to the internet both play a role.

## 1.3 TRANSFORMATION FROM TRADITIONAL CLASSROOMS TO DIGITAL CLASSROOMS

There are various kinds of online learning rooms like blended classrooms (offline & online), flipped classrooms where students study at home and deliver in class whatever they have learned through online information resources (student-centric learning). There are two types of online learning: in one form lectures are recorded by the teacher and made public like in MOOCs (Massive Open Online Courses) and the other form is online classes through Zoom, Google Meet, or other applications. In addition to a reliable IT infrastructure and faculty members who are comfortable teaching online, universities need high-speed internet and education distribution networks or learning management systems. To attend sessions or watch pre-recorded lectures, students may require high-speed internet and computers/mobiles. In India, several networks have been established to facilitate online education. These are supported by the Ministry of Human Resource Development (MHRD), the National Council of Educational Research and Training (NCERT), and the department of technical education. There are also initiatives like:

- e-PG Pathshala (e-content)
- SWAYAM (online courses for teachers)
- NEAT (enhancing employability)
- National Project on Technology Enhanced Learning (NPTEL)
- National Knowledge Network (NKN)
- National Academic Depository (NAD)

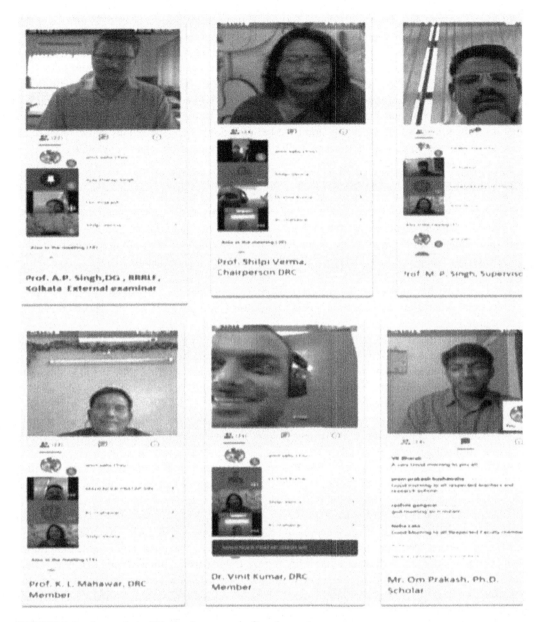

**FIGURE 1.1** Screenshot of Ph.D. viva voce via Google meet.

There are number of platforms available to conduct webinars and online classes such as Zoom, Skype, GoToWebinar, JetWebinar, WebinarNinja, and GoogleMeet.

Figure 1.1 shows the Ph.D viva-voice conducting using Google Meet platform.

The above viva-voice was conducted at the Babasaheb Bhimrao Ambedkar University, Lucknow.

The National Law University of Delhi was one of the first law schools in India to offer an open MOOC, and it did so in March, in the early wake of the COVID-19 crisis. The University Grants Commission (UGC) and the Ministry of Human Resource Development (MHRD) have entrusted to student's research materials in law and digital resources.

In 2003, the National Programme on Technology Enhanced Learning (NPTEL), a project of the Ministry of Human Resource Development (MHRD) that included seven Indian Institutes of Technology (IITs) and the Indian Institute of Science Bangalore, was launched to provide online education. The aim was to create web-based and video-based engineering, scientific, and management courses.

Bharathi Balaji, Head of Operations at NPTEL India, says COVID-19 has pushed institutions, faculty, and students on to online learning like never before. "NPTEL has grown since it was started, but the level of percolation was only decent. In the past six years, we have explained to people what it means to use online education, and tried to break down their inhibitions. Now, because of COVID-19, there is no option but to adapt and utilize online education" (Farooqui, 2020).

## 1.4   FROM PROBLEM TO RESOLUTION

Today, it goes without saying that we should all – schools, universities, local government agencies, non-profits, religious organizations, and others – contribute our time, talent, and resources to the work of teaching, learning, and caring. Although COVID has affected normal life in negatively, but it is not completely. It has also offered some opportunity in following manner:

- COVID-19 has made us more aware of the importance of integrated student support systems (that draw on academic, social, emotional, and physical health data) to serve the whole child.
- COVID-19 made us see and realize that if teachers are to provide such support, they must have more time to work with each other and with helping professionals outside the school.
- COVID-19 helped us to better understand the proper role of online, 24/7 learning environments, spurring innovations for a more accessible, equitable, and personalized system of education from pre-KG to career and college.
- COVID-19 has inspired us to reorganize people and programs in our school districts, universities, and community organizations to work together more effectively to teach, heal, and mentor young people.
- COVID-19 has informed policy leaders as to how government and the private sector can fuel the kind of cross-sector collaboration, from birth to workforce that must be in place to educate everyone in the future (Berry, 2020).

As government has also motivated people to "AAPDA SE AWASAR TAK," the literal meaning being, from disaster to opportunity. It means the government has also encourage people to not take the COVID-19 situation as disaster in their life but try to look from another frame that how it had provided us golden opportunity to achieve another height in life. The old adage, "never let a crisis go to waste," seems apt at the moment. In this time, when things are perilous for so many, we need our policymakers, school reformers, and educators to seek solutions outside the usual places. Students and families clearly need more than our public institutions have been in the habit of providing.

The goal, post-COVID-19, should be to pursue a new kind of community schooling in which:

- Internet access becomes universal for every student.
- Personalized learning is the norm.
- Universities and school districts combine their resources, people, and programs to boost students' college and career readiness and life chances.
- Authentic performance assessments are used at scale to remake accountability and transform the way student and school progress is measured.
- Educators and other helping professionals work together to address the challenges they face and the innovations they seek.

## 1.5 CLASSES IN A MECHANICAL WAY: A CHALLENGE

Online education is fulfilling the demand of classes but at the same time many challenges are also there for the learner and learned. One of the major challenge was that the practical classes are not possible. Sometime face-to-face group discussions and debates in classrooms give solutions for many unanswered queries. These challenges are faced by faculty as well as students alike. Education entails more than just lectures; it entails participation, broadening horizons, free-flowing open debates, and individual student mentoring. If we try to discuss all of these topics in online courses, we will get lost in the process. There is no excitement among teachers as while teaching there is eye contact in live classrooms, but not in online mode as there is no face-to-face engagement with students. The students may be cut off from university library and other resources. Brainstorming exercises, collaborative partnerships, mentoring, and other tasks are possible in live teaching classroom because once you want to do any of this using technology, both the student and the teacher would need to be familiar with it about what and how. Universities are wonderful opportunities for the young people of all backgrounds to connect, collaborate, have fun, and achieve academic goals. Online learning can be used in conjunction with traditional learning.

## 1.6 ONLINE EDUCATION FOR TEACHERS

### 1.6.1 ADVANTAGES

- Allows more flexibility and convenience.
- Allows for creative educational approaches using technology and online resources.
- Provides global reach to a huge number of students.
- It is very beneficial for distance education.

### 1.6.2 DISADVANTAGES

- Very monotonous due to continuous use of laptop or mobile.
- Time and experience are needed to teach online.
- There is consent for fair evaluation of students.
- The inability to engage with students face-to-face facilitates free conversations, exchanges, and mentoring.
- Due to technical constraints, it is impossible to meet all students.

## 1.7 ONLINE EDUCATION FOR STUDENTS

### 1.7.1 ADVANTAGES

- The opportunity to learn through the use of various online resources and processes.
- No interruption in learning due to the COVID crisis.
- Learning at their own pace without any time or space restriction/limitation.

### 1.7.2 DISADVANTAGES

- Disturbance in conversations, debates, and discussions.
- Technological issues or poor access to the internet.
- There are various distractions while learning from home.

## 1.8    POST-COVID IMPACT ON STUDENTS

According to a recent survey by the Organization for Economic Co-ordination and Development conducted of decision makers in 330 educational organizations across 98 countries, including many developing countries, education systems have been facing enormous challenges in addressing students' emotional health. For many students, and especially those living in fragile contexts, school can provide a (relative) haven from violence and other external threats, as well as access to services such as psychosocial support. With the closure of schools, children may be more exposed to gender-based and other violence, including in the home. Other impacts include:

- Students' mental and emotional health will suffer. It is estimated that 10 to 20% of children and adolescents around the world suffered from mental disorders before the pandemic.
- Moreover, research has consistently shown that children often experience psychological stress following natural disasters and other crises.
- School closures, fear of COVID-19, and the social and economic disruptions that accompany the pandemic will likely increase stress within the family and lead to anxiety and depression, including among children and youth. They may suffer fear and grief after experiencing sickness or the loss of friends or family members.
- Research shows that prolonged stress can impair students' learning and threaten their future development.
- Furthermore, parents and teachers may have difficulty responding adequately to threats to students' mental and emotional well-being during the pandemic, given that they lack the necessary training and are likely to experience elevated levels of stress and anxiety themselves.
- Stress among parents is associated with child abuse and neglect, and there is some suggestive evidence that domestic violence has already increased since the COVID-19 crisis began.

## 1.9    KEY PRINCIPLES IN MOVING TO REMOTE LEARNING MODELS

Many education specialists say:

- Children learn in all moments and environments without any hindrances.
- Engage children in online delivery platforms, embracing their capabilities and skills.
- Use multicultural approaches to ensure that all children have equal access to academic opportunities and are fully involved in learning platforms.
- Establish a calm and caring attitude to the unique background and interests of children and their families in order to have a stable influence and emotional well-being for students.
- Educate children and their families about e-safety and being safe while online.

## 1.10    HOW TO IMPROVE LEARNING EXPERIENCES

It is important for educators to maintain a focus on learning outcomes rather than allowing online platforms and resources to govern learning. Learning environments must be structured to accommodate children's learning in a variety of modes.

There are many ways to improve learning experiences:

- Create learning environments that inspire children by allowing them to communicate with their natural surroundings, and then transfer that knowledge to online or paper-based learning.
- Apply flipped classroom and blended learning approaches (a mixture of online and offline activities). That is, allow children to participate in learning events and gather content, objects, and experiments, which they then post online.

- Encourage learners to create investigation questions that they would like to investigate in their natural environment and then share them online with relatives and caregivers.
- Encourage pairs or small groups of children to design a shared inquiry question that they each explore in their own setting and then have children collectively design approaches for sharing their findings.
- Schedule daily and brief online sessions where children can share their learning intentions, discoveries, and learning in a way that accommodates their developmental attention span.
- Design games that combine online experiences with physical movement, such as a scavenger hunt, group stories, singing, or listening to music.
- Establish a daily schedule for using online resources so that children, their friends, and caregivers can get into a routine.
- Integrate a variety of devices that are accessible to young people, such as film, smartphones, and images, as well as voice recordings, digital sketches, and games.
- Allow children to choose from a variety of activities during the week, while maintaining continuity of activities around classroom learning areas – for example, using a rubric to track success.

### 1.10.1 FOR CHILDREN

#### 1.10.1.1 Resources Used for Children Must Be from Natural Settings

Educators provide access to a range of teaching and learning opportunities in early childhood learning. While the transition to remote learning may seem challenging, there are many opportunities for authentic learning experiences. According to Price, "The best learning includes linking authentic, engaging and purposeful household activities with learning experiences – gardening, cooking, painting, building, cleaning, redesigning rooms, and shopping, for example."

Mathematics, English, humanities and social sciences, technology, arts (music, theater, dance, video, and media), fitness and physical education, languages, architecture, and technology are all fields where children's natural habitats are abundant with opportunities that can be extended to all learning areas. This allows children to be challenged to see all of the learning opportunities that exist through their interactions with the natural environment.

For the learning of the children, following natural materials can be utilized as tools and instruments:

> **Household items** – Teaching and learning can be continued using items in the household, such as analogue and digital clocks, a stopwatch, phones, and the oven timer and measuring principles, such as weight, capability of cooking ingredients, liquids, numbers around the building, and numbering of items are all examples of statistical concepts and numeracy.
> **Nature resources** – Consider promoting outdoor play and building using natural materials such as water, bricks, and stones, as well as creating art with nature.
> **Commercial resources** – Some students may have access to a variety of sports and fitness resources such as trampolines, skipping ropes, hoops, and racquets, as well as card, board, and interactive play.
> **Art resources** – Try using homemade play dough, as well as kitchen art or musical instruments (including improvised instruments such as barrels for drums, etc.).
> **Digital resources** – Digital assistants such as Google Home, shared websites, e-books, videos, tablet tablets, phones, laptops, watches, cameras, video games, etc. (Vukovic, 2020).

Two other ideas are to engage children in safe online games (allowing for enjoyable interaction but keeping protection and parental monitoring in mind) and to assign learning activities that include children questioning or finding information from family and friends – allowing them to maintain connections with people outside of their class.

### 1.10.2  FOR ADULTS

During social distancing it is also important for adults to find ways to remain engaged with one another. As a result of physical distancing, during this time of social isolation, creating ways to continue to cultivate social connectedness is critical.

Google Classroom and Zoom are examples of online platforms that allow for whole-class engagement as well as breakout rooms. Both formal group tasks and casual interactions will benefit from break out rooms. Chat forums are also present on some websites for students who do not feel comfortable chatting in a class or community forum and would rather type their conversation. Allowing children to collaborate on the same documents through Google Docs, on the other hand, will encourage connectedness and a mutual purpose.

## 1.11  MONITORING STUDENTS' PROGRESS

"As with face-to-face teaching, assessment can include both formative and summative assessment," Price says, including:

- Observing how children use media tools, how content is presented, and how they react to questions.
- Online/posting/emails of documents, artefacts, images, photos, and videos of learning.
- Teachers can create tests using online quizzes, Mentimeter, Wordle, and other tools.
- Children may use images, drawing, reading, and voice recordings to create journals of their everyday experiences and learning.
- Find out the thoughts and input of your parents, colleagues, and family members.
- Collect the self-reflections and experiences of adolescents.

## 1.12  PARTNERING WITH STAKEHOLDERS

COVID-19 has brought the opportunity for parents and teachers to work more closely together to achieve the best possible learning results and to share responsibilities, respecting the work-culture of the other. The crisis also offers opportunity to build back educational systems stronger and more equitable than before. After the pandemic, parents, teachers, mass media, the government, and others will have changed their views and perceptions about their role in the education process. For example, parents will have a better understanding of the need to work jointly with schools to foster the education of their children. Equity gaps will have been made more evident, along with the urgent need to narrow them. There will be a better understanding of the digital divide – the differences in access to hardware, connectivity, and the right software, but also the huge shortfall of teachers with digital skills. This will create an opening. It is important to use it to build back better. Collaboration has shown what is possible when countries focus on the most effective and equitable approaches to closing learning gaps for all children. It is crucial to learn from those successes and integrate them into regular processes – including through more effective use of technology in remote-learning systems; early warning systems to prevent dropout; pedagogy and curriculum for teaching at the right level and building foundational skills; and ramped-up support for parents, teachers, and students, including socioemotional support.

## 1.13  CONCLUSION

As discussed, the pandemic has affected the education system of the whole world. Its mental impact is visible in students of all age groups. While the learning and practical activities of students have slowed down, as a result of collaboration with education systems and new techniques students have remained motivated. Various types of applications, including online teaching learning platforms, libraries, lectures, and teacher meetings, etc., have supported students in the field of education.

## REFERENCES

Berry, B. (2020). Teaching, learning, and caring in the post-COVID era. Kappan. Teaching, learning, and caring in the post-COVID era – kappanonline.org

Farooqui, S. (2020, May 1). Education in the time of Covid-19: How institutions and students are coping. Business Standards. *Business Standard News.* business-standard.com

Vukovic, R (2020, April 24). COVID-19: Practical tips for early years teaching and learning. COVID-19. www.teachermagazine.com/au_en/articles/covid-19-practical-tips-for-early-years-teaching-and-learning

# 2 Adoption of Blended Learning Model in Higher Education Post-COVID-19

*Tejinderpal Singh[1]\*, Harmeet Kaur Kang[2], and Raj Kumar[3]*
[1]University Business School, Panjab University Chandigarh, India.
[2]Chitkara School of Health Sciences, Chitkara University, Rajpura, Punjab, India
[3]Banaras Hindu University, Varanasi, Uttar Pradesh, India,
\*tejinderubs@gmail.com

## CONTENTS

DOI: 10.1201/9781003358916-2

## 2.1   INTRODUCTION

Most of the educational institutions in India and across the world were closed because of the threatening impact of the COVID-19, an infectious disease caused by a newly discovered coronavirus. According to the daily tracker of UNESCO (May 2020) approximately 1.19 billion learners (68% of total enrolled learners) have been affected due to COVID-19 and subsequently imposed lockdowns in most parts of the world. In India alone, there are 320 million learners who are away from schools and colleges and are temporarily engaged in learning through a wide variety of online platforms. Although Ministry of Education (MOE), Government of India, University Grants Commission, and other academic bodies have issued advisories/guidelines to engage students online through already available platforms of the government or using any third-party platforms However, majority of the teachers are not well acquainted with online learning platforms and tools. Moreover, they have started using online tools just based on their past experience and knowledge only. However, despites many challenges, teachers are making consistent efforts to meet the needs of students in this trying time.

The Health Ministry and other administrative and regulatory bodies have stressed to Indians that we have to learn to live with COVID and social distancing and sanitizing may be a part of our lives for years to come.

However, in the education sector it is challenging to maintain social distancing in classrooms and in other places where teachers and students interact as our infrastructure and other faculties are not designed for social distancing. Moreover, academic calendars are so squeezed that it would not be possible to allow classes at 50% capacity on alternative days or in two shifts.

Morover, we can not solely depend on online education platforms as they are not universally available since parts of India lack of infrastructure and connectivity.

Thus, we need must open our education sector with a new pedagogical approach where it is possible to adjust our classes according to whatever is available to us in the capacity of our available physical and intellectual resources. A new pedagogical approach that will be helpful in this tough time is *blended learning,* which has been in practice for many years but not in the context that we all are presently facing. This chapter is an attempt to highlight the need for the adoption of a blended learning approach by our colleges and universities along with a brief description of various tools and resources teachers can use to make this approach more effective in the post-COVID-19 period.

## 2.2   WHAT IS BLENDED LEARNING?

Blended learning, sometimes called flipped learning, hybrid learning, or technology-mediated instruction, is the combination of traditional classroom teaching with online educational materials and methods. In this approach both teacher and student are physically present in a brick-and-mortar system and interact with each other face to face. However, some content is offered online to students where they have control over time, place, or pace regarding the consumption of the content. Let us look at a few scenarios of blended learning.

**Scenario 1:** *As a teacher, my first class with MBA students is to explain the concept of research. I will ask students to watch three videos (URLs provided) before coming to the class and prepare a note on which one out of three videos is an example of research and why. During my face-to-face interaction in the classroom I will listen to the students' answers and explain the meaning of research by linking it with the videos they watched. Here, I am using online and offline platforms in my teaching, which is an example of blended learning.*

**Scenario 2:** *I have a class on the topics of 'search engine optimization' that simply means bringing your website to the top of Google search results. Now, to understand this topic well students should first have knowledge about 'how Google Search Engine works' and 'how* *websites are developed'. I ask my students to watch online my pre-recorded lectures on 'How Google works' and 'website development' before coming to class. This way students have a fair idea about the building blocks of search engine optimization when they come to my class.*

**Scenario 3:** *I want to teach my students on how to use Facebook to promote products and services. As it requires demonstration of the process to start online advertising, I record my screen with my narration and develop a complete lecture on this process, I share it with my students and ask them to practice at their own end. In this case, I do not call all the students to my class. Only half of the students come to me for more details on the topic or for any doubt clearing; the rest of the students come on the next day for the same purpose. This way, there is a possibility to maintain social distance and other safety norms with half of the students in the class. In this way students do not feel disconnected from the teacher and their fellows students.*

From these examples, it is clear that some aspects of a course can be covered online to give students more flexibility to go through classroom material and save time for teachers that can be used to organize discussion in the classroom. Case 3 will be scenarios of our classes in the days to come under the blended learning approach.

## 2.3   WHY BLENDED LEARNING?

The effectiveness of blended learning has been empirically studied by various scholars (López-Pérez et al. 2011; Means et al. 2009; Garrison and Kanuka 2004). It offers the necessary tools to engage learners and to increase instructor productivity. The following points highlight the significance of blended learning.

### 2.3.1   LEARNING HAPPENS ANYWHERE, ANYTIME

In the present scenario, a teacher cannot be present all the time in a classroom. Some portion of the course work is delivered online such as short videos, cases, problem-solving exercises, group discussions, and so on. Similarly, all the students cannot be present in the classroom all at once to digest the content offered by a teacher. Since blended learning content is offered online, it provides flexibility to the learners to learn when and where they are ready using their laptops, tablets, or mobile phones.

### 2.3.2   INCREASED ENGAGEMENT

According to Bradbury (2016) the attention span of students ranges from 10 to 15 minutes during lectures. It is a very challenging task to actively engage learners for longer durations, especially in disciplines like management. A blended learning approach offers a series of activities, interaction, feedback, and quizzes in a way that otherwise may not be possible in a purely face-to-face learning environment. While completing these activities online students are automatically engaged in the learning process.

### 2.3.3   CONVENIENCE

With the advent of technology in the education field, there are many tools available for teachers that allow them to upload their course material online, generate quizzes, monitor student performance, accept assignments, and give feedback in an online platform.

### 2.3.4 Multiple Tools

A blended learning approach allows the use of a variety of tools for communication, interaction, monitoring, testing, evaluation, gamification, and grading such as Google Classroom, Moodle, Blackboard, etc. Material uploaded on these platforms can also be reused for subsequent sessions without issue.

### 2.3.5 Easy Student Feedback

Blended learning also offers the ability to rapidly analyze, review, and give feedback on student work. It also allows teachers the ability to tailor teaching methods and feedback to each student while improving time efficiency.

## 2.4 MODELS OF BLENDED LEARNING

The literature reveals that there is little consensus on the definition of blended learning. Researchers and academicians have proposed various models of blended learning. Friesen (2012) classified blended learning into six different models as follows.

### 2.4.1 Face-to-Face Driver

In this model of blended learning a teacher delivers his lectures to students augmented by digital tools such as online videos.

### 2.4.2 Rotation

In this form of blended learning a student moves from one platform of learning to another and one of the platforms is online. For example: A management student has to attend the teacher's class, has to visit an organization, has to go to the field to interact with customers, and has to complete some class assignments online as well.

### 2.4.3 Flex

In this model, most of the course content is delivered online and teachers are present for face-to-face interaction as well.

### 2.4.4 Labs

All course-related content is delivered in an online platform but in a consistent physical location.

### 2.4.5 Self-Blend

In this model, students can augment their traditional learning with digital content or online courses.

### 2.4.6 Online Driver

In this model, students are supposed to complete their entire course work online. However, there are possible checks of the teachers as well and some face-to-face interaction may happen.

## 2.5 DESIGNING BLENDED LEARNING COURSES

As discussed earlier, blended learning offers numerous advantages to teachers and students. But the majority of the teachers either are not aware of blended learning pedagogy or are not using it to its full extent. Here are a few guidelines extracted from the literature to develop a successful blended learning course plan (Kerres and Witt 2003; Garrison and Kanuka 2004; Friesen 2012; Boyle et al. 2004).

### 2.5.1 UNDERSTAND YOUR CONTEXT

Every teacher's course is different, having a variety of challenges. First and the foremost important is to understand the context in which you are going to design your blended learning course. For example, a teacher offering course in digital marketing has different requirements to blend his content with an online platform than a teacher of music or physics.

### 2.5.2 LOCAL CHALLENGES MATTERS

India is a land of diversity and it may pose technological, socioeconomic, or cultural challenges to the adopters of blended learning. All learning groups are not the same as their access to online content may be different or there can be infrastructural issues like slow internet. Therefore, the content delivery system may not be uniform across all learner groups. A teacher has to make sure that the way he wants to blend the course is in line with the requirements of the targeted learning group.

### 2.5.3 MAP OUT SHORT-TERM AND LONG-TERM GOALS OF BLENDED LEARNING

While designing a blended learning course, map out the expected outcomes of the course. Divide the entire course into various modules and then set learning objectives for each module. Now for each module, determine which part will be discussed in the class and which part students will cover online. For each module, determine how students will be assessed and what activities are expected from them. Table 2.1 shows an example of a blended learning module template.

### 2.5.4 MAKE ONLINE LEARNING COMPLEMENTARY TO CLASSROOM LEARNING

Online learning is an essential component of blended learning. A teacher can develop his own |e-content or ask students to join online courses at third-party platforms to complement the classroom teaching. For example, the Indian Institute Management Bangalore offers management courses on its IIMBx Platform to management students and SWAYAM offers free online courses in

---

**TABLE 2.1**
**Course Plan Blended Learning Template**

| Name of Module | Learning Outcome | Mode of Learning | Activities | Assessment |
|---|---|---|---|---|
| An Overview of Business Research | -To Comprehend the Meaning of Research<br>-To Explain, How Academic Research is Different from Professional Research? | F2F (Face to Face), Online Discussion, and Google Classroom | 1 and 2* | Online Quiz. Available at www.singhtejinder.com/ ubsmoodle |

\* Activity-wise detail should be provided to the students.

*Source*: Adapted from course plan on qualitative research methods of the author.

**TABLE 2.2**
**List of MOOC Platforms**

| Sr. No. | Organization | URL |
|---|---|---|
| 1. | IIM Bangalore | www.iimb.ac.in/iimbx |
| 2. | Edex | www.edx.org/school/iimbx |
| 3. | Swayam | swayam.gov.in/ |
| 4. | Udemy | www.udemy.com |
| 5. | Khan Academy | www.khanacademy.org/ |
| 6. | Coursera | www.coursera.org/ |

*Source*:   Compiled from various sources (newspaper reports, etc.).

**FIGURE 2.1**   Classification of blended learning tools.

all disciplines with a facility of credit transfer. Table 2.2 provides a list of various online platforms for MOOC (Massive Open Online Course) students can join as a part of blended learning.

### 2.5.5 USER FEEDBACK IS IMPORTANT

Paying attention to the feedback received from learners when developing blended learning courses is important to develop the course as per the requirements of actual users. Various online tools may be used such as Google Form, Survey Monkey, Kobo ToolBox, etc., to gather student feedback.

## 2.6 BLENDED LEARNING TOOLS

A variety of blended learning tools is available to teachers to help them to adopt the blended learning model. Blended learning tools can be classified into five categories based on: Learning management systems, E-content development, interactive broadcasting, student engagement and student assessment. Let us briefly learn about these tools (Figure 2.1).

### 2.6.1 LEARNING MANAGEMENT SYSTEMS

A Learning Management System, or LMS, is the backbone of any online educational programme including blended learning approaches. A LMS is a cloud-based online platform that allows teachers to organize their course content in the form of classes. Actually, it is just a shift of the physical class to the virtual world where a teacher and students interact with each other to make the teaching learning process more effective and impactful. The process of using an LMS is simple.

The teacher will create a class on a specific LMS such as Google Classroom and the respective students will register for that class either using a unique class code or through invitation from the teacher. Once a class is created a teacher can add the material and resources to the class, assign activities to students, conduct online quizzes, evaluate assignments, make comments on assignments, and create a gradebook for each of the students. Similarly, through an LMS a student can access

**TABLE 2.3**
**A List of Popular Learning Management Systems**

| Sr. No | Name | Links |
|--------|------|-------|
| 1. | Google Classroom | https://classroom.google.com/ |
| 2. | Moodle | https://moodle.org/ |
| 3. | Blackboard | www.blackboard.com/index.html |
| 4. | Edmodo | www.edmodo.com/ |
| 5. | Talent LMS | www.talentlms.com/ |
| 6. | Easy LMS | www.easy-lms.com/ |
| 7. | Learn Dash | www.learndash.com/ |
| 8. | Canvas | www.instructure.com/canvas/en-au |
| 9. | Mind Scroll | mindscroll.com/ |

*Source*:   Complied by the authors.

teacher material and resources, submit assignments, ask questions, initiate discussions, view the comments on assignments, and check grades.

Table 2.3 lists few popular LMSs along with their links.

### 2.6.2   E-CONTENT DEVELOPMENT TOOLS

The blended learning approach requires every teacher to develop e-content to be shared with students in a digital format. E-content generally includes text, videos, audio, images, and animations. E-content may be a single element like video or audio or it may include more than one element such as video, PPT, audio together in a single webpage. Teachers must learn to create e-content or to convert already existing content to e-content form. Table 2.4 lists some of important tools for developing e-content.

### 2.6.3   INTERACTIVE BROADCASTING

In a blended learning approach, interactive broadcasting of class lectures is sometimes used to reach students, especially in times of crisis, such as during COVID-19. During interactive broadcasting, a teacher delivers the lectures live using any online platform where students are virtually connected with the teacher on the same platform. During the session, the teacher and students are face to face and may ask any query during the session that can be resolved by the teacher in real time. During the session, a teacher can get quick feedback from the students about their understanding of the topic. There are many free and paid tools available for teachers to organize an interactive broadcasting event as shown in Table 2.5.

### 2.6.4   STUDENT ENGAGEMENT

Student engagement is the major concern both in online teaching as well as in blended learning. In traditional classrooms, students are more controlled by teachers as students and teachers are face to face. However, when e-content is shared with students in a blending learning approach, it is very difficult to monitor students, to see if a student is reading the content or watching the videos or visiting the websites. If students are not actively participating in the class or not showing interest in

**TABLE 2.4**
**E-Content Development Tools**

| Sr. No. | Form of E-content | Name of the Tool | Site Link |
|---|---|---|---|
| 1. | Video Creation | OBS (Open Broadcaster Software) | https://obsproject.com/ |
| | | Loom | www.loom.com/screen-recorder |
| | | Icecream Screen Recorder | icecreamapps.com/Screen-Recorder/ |
| | | Screencast-O-Matic | https://screencast-o-matic.com/ |
| | | Camtasia | www.techsmith.com/video-editor.html |
| | | Screencastify | www.screencastify.com/ |
| 2. | Text Creation | Google Docs | docs.google.com/ |
| | | Google Slides | www.google.com/slides/ |
| | | Office 365 | www.office.com/ |
| 3. | Animation Creator | Blender | www.blender.org/ |
| | | Animaker | www.animaker.com/ |
| | | Moovly | www.moovly.com/ |
| | | VYOND | www.vyond.com/ |
| | | PowToon | www.powtoon.com/ |
| | | Doodly | www.doodly.com/ |
| 4. | Audio Tools | Vocaroo | vocaroo.com/ |
| | | Chirbit | www.chirbit.com/ |
| | | Audacity | www.audacityteam.org/ |
| 5. | Images | Pixabay | https://pixabay.com/ |
| | | Unsplash | https://unsplash.com/ |
| | | Canva | www.canva.com/ |
| | | ShutterStock | www.shutterstock.com/ |

*Source*:   Complied by the authors.

**TABLE 2.5**
**Tools for Online Interactive Broadcasting**

| Sr. No. | Name the Tool | Site Link |
|---|---|---|
| 1. | Google Meet | https://meet.google.com/ |
| 2. | Facebook Live | www.facebook.com/ |
| 3. | Zoom | https://zoom.us/ |
| 4. | GoToMeeting | www.gotomeeting.com/ |
| 5. | BigBlueButton | https://bigbluebutton.org/ |

*Source*:   Complied by the authors.

the content shared, revisions need to be made to make the blended learning approach effective and productive. A teacher should always make an attempt to engage students by developing content that is innovative and interesting. Table 2.6 lists some tools teachers can use to engage students and to make learning a fun activity.

### 2.6.5   STUDENT ASSESSMENT

Students assessment, a process of gathering and evaluating the gaps between knowledge disseminated and knowledge retained, is an integral part of any education program. In the traditional assessment

**TABLE 2.6**
**Digital Tools for Student Engagement**

| Sr. No. | Name the Tool | Site Link |
| --- | --- | --- |
| 1. | Insert Learning | https://insertlearning.com/ |
| 2. | Edpuzzel | https://edpuzzle.com/ |
| 3. | Kahoot | https://kahoot.com/ |
| 4. | FlipGrid | https://flipgrid.com/ |
| 5. | AnswerGarden | https://answergarden.ch/ |
| 6. | Coggle | https://coggle.it/ |
| 7. | Padlet | padlet.com/ |
| 8. | Remind | www.remind.com/ |
| 9. | PearDeck | www.peardeck.com/ |
| 10. | Buncee | app.edu.buncee.com/ |

*Source*: Complied by the authors.

**TABLE 2.7**
**Student Assessment Tools**

| Sr. No. | Name of the Tool | URL |
| --- | --- | --- |
| 1. | Google Form | www.google.com/forms |
| 2. | Formative | https://goformative.com/ |
| 3. | Mentimeter | www.mentimeter.com/ |
| 4. | Micropool | www.micropoll.com/ |
| 5. | Quizlet | quizlet.com/ |
| 6. | Socrative | www.socrative.com/ |
| 7. | Quizalize | www.quizalize.com/ |
| 8. | Peergrade | www.quizalize.com/ |
| 9. | AnswerGarden | https://answergarden.ch/ |

*Source*: Complied by the authors.

model, teachers evaluate students through an in-person proctored examination system. In a blended learning approach, some portion of overall student assessment should be online. While this may appear to be challenging, there are multiple online tools available that can be used to assess students as shown in Table 2.7.

## 2.7 GOVERNMENT INITIATIVES TO POPULARIZE BLENDED LEARNING

The Government of India has been making consistent efforts to promote e-learning/blended learning in the country especially during COVID-19. The launching of the SWAYAM (Study Webs of Active-Learning for Young Aspiring Minds) platform is a great initiative in this direction. SWAYAM offers online courses ranging into the thousands – mostly taught at high school, college, and university level and is free of cost.

The main aim of the SWAYAM programme is to achieve the three cardinal principles of education policy: access, equity, and quality. It also strives to offer the best teaching learning resources and material to all, including the most disadvantaged.

**TABLE 2.8**
**List of SWAYAM National Coordinators**

1. AICTE (All India Council for Technical Education) for self-paced and international courses
2. NPTEL (National Programme on Technology Enhanced Learning) for engineering
3. UGC (University Grants Commission) for nontechnical post-graduation education
4. CEC (Consortium for Educational Communication) for under-graduate education
5. NCERT (National Council of Educational Research and Training) for school education
6. NIOS (National Institute of Open Schooling) for school education
7. IGNOU (Indira Gandhi National Open University) for out-of-school students
8. IIMB (Indian Institute of Management, Bangalore) for management studies
9. NITTTR (National Institute of Technical Teachers Training and Research) for teacher training programme

*Source*:    Swayam Central 2019.

**TABLE 2.9**
**List of Blended Learning Resources along with Links**

| | | |
|---|---|---|
| UG/ PG MOOCs | All SWAYM courses are archived on this platform. Students can access the content of these courses any time. | ugcmoocs.inflibnet.ac.in/ugcmoocs/ moocs_courses.php |
| e-PG Pathshala | More than 23,000 curriculum-based modules are available in 70 postgraduate disciplines | https://epgp.inflibnet.ac.in/ |
| e-Content Courseware in UG Subjects | E-content of more that 24,000 modules is available in 87 undergraduate subjects | cec.nic.in/ |
| SWAYAMPRABHA | A group of 32 DTH channels providing high-quality educational content in diverse disciplines to teachers and students across India | https://swayamprabha.gov.in/ |
| CEC-UGC YouTube Channel | Unlimited number of curriculum-based educational lectures are available | www.youtube.com/user/cecedusat |
| National Digital Library | A digital repository of academic content in various forms | https://ndl.iitkgp.ac.in/ |
| Sodhganga | A digital repository of 260,000 Indian thesis and dissertations | https://shodhganga.inflibnet.ac.in/ |
| e-Shodh Sindhu | Access to more than 15,000 peer-reviewed journals | https://ess.inflibnet.ac.in/ |

SWAYAM additionally looks to connect those who remain untouched by the digital revolution and are not able to join the mainstream of knowledge by closing the digital divide.

The courses available on SWAYAM are in four quadrants: (1) video lecture; (2) downloadable text in PDF form; (3) self-assessment tests through tests and quizzes; and (4) an online discussion forum for getting questions answered. At present, nine national coordinators have been appointed to ensure the quality of online courses. These nine coordinators are shown in Table 2.8.

Further, MOE and UGC jointly have taken many ICT initiatives to actively engage students in online learning. All the listed resources can be utilized by teachers in the blended learning approach as these resources are free to use (Table 2.9).

## 2.8 CONCLUSION

There is no doubt that teaching can be more effective and engaging if a blended learning teaching pedagogy is adopted, especially post COVID-19. However, this approach may have a few disadvantages since it has a strong dependence on technology and the internet. There can also be issues in the implementation if the target students are from rural or remote areas where internet connectivity and bandwidth is still a problem. Moreover, not all teachers are well conversant with the use of technology in teaching, which may hinder the process of adoption of blended learning. Therefore, it is the need of the hour to organize special hands-on training programs or workshops to make educators familiar with the basic tools of online teaching. Although digital literacy is low in India and can be a significant barrier in this regard, if blended learning is promoted among the teaching fraternity, better teaching learning experiences especially during stressful times can result.

## REFERENCES

Boyle, Tom, Claire Bradley, Peter Chalk, Ray Jones, and Poppy Pickard. (2004). Using Blended Learning to Improve Student Success Rates in Learning to Program. *Journal of Educational Media. 28*(2–3): 165–178.

Bradbury, Neil A. (2016). Attention Span during Lectures: 8 Seconds, 10 Minutes, or More? *Advances in Physiology Education. 40*(3): 509–513.

Dhawal, S. (2018). By the Numbers: MOOCs in 2018-Class Central. www.classcentral.com/report/mooc-stats-2018/.

Friesen, N. (2012). Report: Defining Blended Learning. *Norm Friesen,* August. Available at www.normfriesen. info/papers/ Defining_Blended_Learning_NF.pdf

Garrison, D. Randy, and Heather Kanuka. (2004). Blended Learning: Uncovering Its Transformative Potential in Higher Education. *Internet and Higher Education.* 7(2): 95–105

Kerres, Michael, and Claudia De Witt. (2003). A Didactical Framework for the Design of Blended Learning Arrangements. *Journal of Educational Media. 28*(2–3): 101–113.

López-Pérez, M. Victoria, M. Carmen Pérez-López, and Lázaro Rodríguez-Ariza. (2011). Blended Learning in Higher Education: Students' Perceptions and Their Relation to Outcomes. *Computers and Education. 56*(3)1: 818–826.

Means, B., Y. Toyama, R. Murphy, M. Bakia, and K. Jones. (2009). Evaluation of Evidence-based Practices in Online Learning: A Meta-analysis and Review of Online Learning Studies. Available at www2.ed.gov/rschstat/eval/tech/evidence-based-practices/finalreport.pdf

Statista (2016). Statistics Portal 2016. Available at www. statista.com/statistic-portal/.

Swayam Central (2019). Available at https://swayam.gov.in/about.

TRAI (2016). Telecom Subscriptions Report. Available at www. trai.gov.in/sites/default/files/Telecom Sub_Eng_pr.03_09-01-2017.pdf.

UNESCO (2020). Available at https://en.unesco.org/covid19/education response

# 3 ODE

## Cruising onto a New Education System

*Ravi K. Mahajan[1] and Kalpana K. Mahajan[2]*
[1]USOL, Panjab University, Chandigarh, India
[2]Department of Statistics, Panjab University, Chandigarh, India

## CONTENTS

## 3.1 NINETEENTH CENTURY: ERA OF GERMINATION

The modern education system owes a lot to the French Revolution, which unraveled to commoners the colossal importance of education. In the vibrancy of the post-revolution era, some enthusiasts took the central stage to share their expertise/skills with others at the individual level, and others institutionalized new educational avenues. For sure, the nineteenth century saw the emergence of terms like non-formal, informal, and formal to address endeavors for the enrichment and empowerment of learners. Beneath these pursuits was the power of 'Ekalavyaism', which provided an unparallel instance of self-learning with a meditative mind without the physical presence of a 'guru'. Caleb Phillips' experiment of 1728, offering learning opportunities to aspirants without the physical presence of a teacher, was regenerated in the nineteenth century in the ventures by the British Isaac Pitman, the Frenchman Charles Toussaint, and the German Gustav Langencheidt, as well as by the University of London (Anand, 1979; Bhatnagar, 1997). The University of London, founded by a Royal Charter on 28 November 1836, merely focused on registering and examining students in the UK and overseas (History of the University of London, 2020). The university required its students to 'prepare of their own', without any support services. Towing the lines of the University of London, in the nineteenth century a number of universities including University of Calcutta, University of Madras, University of Mumbai, and University of the Cape of Good were established (Anand, 1979).

Nonetheless, in 1840, in Britain, with the introduction of the first postage stamp, Sir Isaac Pitman started a correspondence course by snail mail enabling students to learn shorthand. Pitman, usually said to be the 'father of distance learning', had in fact stepped ahead of University of London, by providing 'student support' in the form of notes in shorthand to learners that they could used to study at their own 'place and pace'. Around the same time, in 1856, the Frenchman Charles Toussaint and the German Gustav Langencheidt formed and organised a correspondence school in Berlin in Germany (Holmberg, 1995). While Toussaint and Langencheidt focused on modern languages, Dr. William Rainey Harper – often referred to as 'the father of American correspondence study' – initiated the Correspondence School of Hebrew in 1881, and HS Hermod in Sweden began teaching English by correspondence in 1886 (Harper-Hermod, 1881–6).

Promotion of languages apart, towards women empowerment, Anna Eliot Ticknor founded in 1873 the 'Society to Encourage Study at Home' and ran it from her home in Boston in the United States. The Society functioned independently and had no affiliation with any college and university. And, for that reason, distance education scholars called it 'The Silent University'. Some other dimensions to the applications of the non-traditional modus operandi were added with the establishment of Skerry's College in Edinburgh facilitating the preparation of candidates for the civil service examinations, Foulks Lynch Correspondence Tuition Service for specializing in accountancy, Tokyo SenmonGakko offering correspondence courses in politics and economics, Moody Bible Institute in the United States, and Thomas J. Foster's pamphlets on 'accident prevention for mine workers' (Pioneers, 1800).

This non-traditional modus operandi of education has been given names such as correspondence courses, external studies, extension education, off-campus, non-traditional, independent learning, home studies, individualized learning, and open learning, to name a few (Anand, 1979; Brassard, 1982; Moore, 2001; Mahajan, 2011). Notably, the term 'correspondence study' was first proposed by the University of Wisconsin in its (1889–1890) catalogue as a function of university extension (Growinstis, 1900). Nomenclatures notwithstanding, the focus point of all these endeavors has been the zeal to reach learners who for a multiplicity of reasons could not derive the benefits of a formal education system. The long array of terminologies wontedly addressed as 'correspondence education' was metamorphosed into what is arguably the most pervasive, most talked about, and most globally recognized form of instruction at a distance (Jegede & Naidu, 1999).

Growing terminologies apart, the era also witnessed some technological contributions that re-engineered the world order. For educationists, the creativity and innovations of Charles Babbage, Samuel Morse, Charles Thurber, and Thomas Edison are manifested in the efficacy and effectiveness of the open & distance education (ODE) system (Tech Contributors, 1880). Interestingly, some visionary statements made in the context of ODE in the nineteenth century seem quite contemporary. For instance, in 1880, William Rainey Harper, one of the leading American academic leaders who helped establish both the University of Chicago and Bradley University, said, 'The correspondence teacher must be painstaking, patient, sympathetic and alive; … whatever a dead teacher may accomplish in the classroom, he can do nothing by correspondence'. And five years later, he said, '… innovations in transportation and postal delivery meant that … the day is coming when the work done by correspondence will be greater in amount than that in the classrooms of our academies and colleges' (Keegan, 1986).

## 3.2   THE TWENTIETH CENTURY: THE ERA OF FRUCTIFICATION

The outset of 20th century was marked by an affirmative note on the increasing scope and growing enrollment in correspondence courses. In the United States, ICS shared that 'by the first decade of the twentieth century, over 100,000 new students per year were enrolling in ICS courses; by 1910, a million cumulative enrollments had been achieved; and, by 1930, four million' (International Correspondence Schools of Scranton, Pennsylvania – 1891 to the present). Gradually, this non-formal stream of education emerged even stronger by showcasing its potential in empowering and enriching aspirants hailing from special segments worldwide. In the United States, the American Institute of Banking formed the Correspondence School of Banking, and Moody Bible Institute established a correspondence school. In Canada, the University of Saskatchewan started providing off-campus learning opportunities such as 'Better Farming' demonstration training, the 'Homemaker' short course, and 'Canadian Youth Vocational Training Workshop'. In India, Rabindranath Tagore's 'Lok Sikkha Samsad', 'Jamia Urdu' at Agra gave an initial boost to this scheme of education (Growinstis, 1900).

So intense was the sanguinity about this upcoming stream of education that Thomas Edison, himself an author of an ICS course in 1910, had remarked, 'Home study is one of the greatest inventions of the twentieth century' (Odesayings, 1938). Technological advancements fructified in radio, television, and computers indeed played a crucial role in the horizontal and vertical growth of this nonformal stream of education during the nineteenth century. In the United States, while the number of colleges and universities having granted radio broadcasting licenses crossed 200 in 1925, the State University of Iowa started offering course credit for five radio broadcast courses. In this context, the role of the BBC in setting up its own adult education section needs a due reference. The vision of 'wireless university' of J.C. Stobart, the Director of Education at the BBC, still deserves emulation. In the United States, the Federal Radio Commission established by the Radio Act of 1927 had also put forth a precise statement on the potential of radio for educational purposes. In India, where radio made its formal entry in the public domain in 1927 at Bombay and Calcutta, education was first taken up by radio in 1929 in Bombay on an occasional basis. In the following year Sri Lanka and Japan also started using radio for education (Radio, 1920s).

By the time radio had virtually stamped its role in the education, television started substantiating its potential in education. Purdue University in 1931 developed the experimental television teaching program, and the University of Iowa in 1933 began televised course broadcasts in various subjects including oral hygiene and identifying star constellations. The introduction of cable television in 1948 as Community Access Television (CATV) in Pennsylvania was the hallmark of the first half of the 20th century. Interestingly, by the time television had inscribed its integral role in education, computers' unmatchable ascend took the stage. Within a decade, computers started playing an indispensable role in lifestyles worldwide. In the education system computers assumed an all-pervasive role, from 'Desktop Publishing', and 'Data Management', to 'Delivery in Teaching-Learning Process'. The use of computers got increased even more with satellites weaving webs on internet (Mahajan, 1993). With respect to technological advancement in the context of education and more precisely distance education, while Tony Bates and O. Peters had seen the growth through 'three stages', Taylor suggested 'five generations of distance education' (Thinkers, 1900).

Amidst the growing interface of technology in education, debates questioning the role of audiovisuals in education have also been echoing in some quarters. An earlier noteworthy reference could be seen in the year 1927. In 1927, during the 'first Republic in Portugal', the possible advantages and dangers of the use of audiovisual aids in the educational process emerged as an issue. Five years after, due to the assumed importance of cinematography in educating people, a commission was formed the called Comissão do Cinema Educativo (Commission for Educational Cinema) under the Ministry of Public Instruction, with the objective of proposing the production, authorship, and distribution of educational films (Thinkers, 1900).

Amidst all permeating expansion in the realm of ODE, the century was also marked by candid concern of stakeholders in promoting and safeguarding this stream of education. A number of associations, newsletters, journals, and conferences dedicated to the spirit of ODE started propagating the thoughts of a new rank of theorists and practitioners. For instance, in 1914, in China, the first Corresponding Association was set up by the Commercial Printing House, and the all-important, International Council for Correspondence Education was founded in 1938. The spirit of the mission was well reflected in the opening address at ICCE's 1st International Conference by Dr K.O. Brady, who said, 'By equality of educational opportunity we mean extending education of equal quality to everyone, no matter how humble his birth, no matter where he may live, and no matter what his reasonable aspirations may be…' (Odesayings, 1938).

In the second half of the century, an array of distance educators like O. Peters, G. Dohmen, Michael G. Moore, Desmond Keegan, Charles A. Wedemeyer, Börje Holmberg, Tony Bates, John Baath, and John Daniel spearheaded the move to provide a strong theoretical basis to open and distance education (Thinkers, 1900).

Their attempts paved the way for ODE to become an independent area of study and even that of a faculty in some universities. However, unmindful of the developments and growing role and importance of ODE, a section of traditional academia continued discounting ODE and even described distance education as 'no more than a hodgepodge of ideas and practices taken from traditional classroom settings and imposed on learners who just happen to be separated physically from an instructor' (Odesayings, 1938).

ODE has met the challenge of providing educational avenues to the exponentially growing population in the backdrop of the expanding horizon of the information age and resource constraints. Yet some exponents of formal systems, particularly the 'non-contributors to ODE', take every opportunity to criticize new systems of learning. There are a variety of reasons for this. Ironically, not uncommon were/are instances where persons who were/are called upon to work on the policy initiatives on ODE or given to hold key positions in the institutes of ODE, with little understanding of the concept, philosophy, and constraints of ODE. While some enthusiasts have seen ODE as a way to 'take their institutes/universities out of red', this form of education was originally developed to provide, with adequate support services, a second chance for enhancement or improvement of academic qualifications to those who, for various reasons, could not do join formal systems of education (Mahajan, 2012).

Nonetheless, the decade ended with a note on quality consciousness and establishment of bodies to oversee the quality and standards in ODE at the global level.

## 3.3    THE TWENTY-FIRST CENTURY: A NEW SYSTEM IN THE MAKING

The agenda for the new millennium was set with the launch of the Global Development Learning Network (GDLN) on 21st June 2000. At its launch, James D. Wolfensohn through videoconferencing was joined by leaders from fourteen countries at participating distance learning sites for a discussion on the benefits of leading-edge tool to bridge the 'digital divide'. The references to the terms 'videoconferencing', 'distance learning sites', and 'digital divide' speak volumes about the new millennium's 'communication system', 'modus operandi', and the 'target', the underlined key words for the twenty-first century. Decisive actions towards attaining the holistic goals could be seen in the launch of a number of initiatives set in motion in very first year of the third millennium. For instance, while the Singapore e-Learning Framework (SeLF) was set up to provide a basis for the e-learning industry, Cyber University of Korea (CUK) was established as the first cyberuniversity in Korea. In Myanmar, the University Correspondence Courses launched e-Education by using a satellite data broadcasting system. In the United States, Illinois Virtual High School began offering instruction. In India, the Government of Tamil Nadu established the Tamil Virtual University (TVU) as a society to develop and deliver Internet-based learning material in Tamil language, literature, and culture (Growinstis, 2001).

The outset of the new millennium was also marked by some initiatives by world bodies like the World Bank, UNO, and Commonwealth of Learning to harness ODE address their respective goals. For instance, the World Bank provided a US$2.0 million credit to Sri Lanka for a distance learning project to create specialized training opportunities for senior decision-makers. In Canada, towards quality standards for online learning products and services, CoL initiated work with the Canadian Association for Community Education (CACE), FuturEd Consulting Education Futurists (FuturEd), and other partners. In this array, the endorsement of UN General Assembly for holding of the World Summit on the Information Society (WSIS) cannot be undermined (Grow Instis, 2001).

In the backdrop of all these information and communication technology/technologies (ICT), the new millennium also saw the emergence of some 'content management systems' and 'learning management systems'. These systems, such as, ATutor, Chamilo and Moodle, to name a few, facilitate the process of administration, documentation, tracking, reporting, automation, and delivery of educational courses.

The increasing concern for open, flexible, accessible, and inclusive quality education found manifestations in three new millennium terms: OpenCourse Ware (OCW), Open Educational Resources (OER), and Massive Open Online Courses (MOOCs). These three terms were coined in the first decade of the new millennium (Terms, 2000).

Interestingly, the march towards 'openness' in education has a long history. It has seen shades in interpretations and adaptations; from a model focusing on 'open entry' to 'open exit', and then to 'open access to content and resources'. It would be no exaggeration to say that the 'open' movement is as old as the very concept of education. Nonetheless, the 'openness embedded' terms OCW, OER, and MOOCs have in fact unleashed a new life in the academic world. As a result, academia became sharply divided between 'formal/regular/face-to-face' and 'open and distance education', which now stands compounded on one platform. Distance educators are pleased that they have another way to attain their aims and objective, and proponents of 'formal/regular/face-to-face'- systems are happy to have a mode to share their expertise with masses at the click of a mouse.

Interestingly, the debates for and against OCW/OER/MOOCs were lost during the pandemic. Courtesy of Covid, educational systems and subsystems world over are being reengineered towards a new world order.

## 3.4 EMERGING MODEL

COVID-19, which brought the world to a standstill in the beginning, soon became affiliated with terms like virtual classrooms, teleconferencing, webinars, web-workshops, web talks, etc. These new terms virtually inverted the pyramid for classrooms and academic events like conferences, seminars, symposiums, and workshops. These academic events, where each event has its own slant with an array of benefits and drawbacks, are integral to the growth of information and knowledge of academic disciplines.

In the pre-COVID-era, to share thoughts and expertise, teachers/researchers/scholars/experts used to travel to preannounced venues, at times even as distant at thousands of kilometers. With web academic events, experts are emerging at the global network at a click on the mouse. The expertise of a genius is no longer confined to a local audience. Newer thoughts are no longer constrained by geographic proximities of an expert or by the paucity of funds. Interestingly, all this is achieved in the most optimized way – saving time and money. No travelling is required, thus saving on fuel, and hence curbing environmental pollution. No preconference meetings. No big banners. No decorations. No wastage of papers on invitation cards, registration forms, schedules, and proceedings. No lodging pressures. No momentous or gifts. No teas or high teas. No loose talks. No thoughtless or rushed presentations. Now these events are all professional, with a focus on content and enrichment (Mahajan, 2020). Much of the funds saved by going online can be used for the improvement of education boosting the teaching-learning process or providing extra help to remote learners.

As in the case of conferences, Covid has also changed classroom styles. While teaching was once characterized by one teacher addressing a class packed with students with minimal time for enrichment, with classes going online each student feels like he or she is getting one-on-one time with the teacher.

Of course, in the array of limitations in 'going virtual', issues like 'undertaking practical', 'personal and tender touch' or 'fragrance of the fellow students and teachers', or the like clamor high. Indeed, concerns haunting 'conduct of practical' and 'remotely placed students' could be worked out by special strategies. Blended learning environment could be easily invoked. A set of preannounced sessions for practical and problem solving, and providing study material in loaded 'tabs', could be articulated in the array of solutions.

## 3.5 CONCLUSION

COVID-19 has successfully dispelled the myth that learning can only happen in a classroom. Now the meeting point for a class is an unbounded virtual space, where transparency is the hallmark of the emerging new teaching learning model. The pressure of being in the public domain, of course, will make teachers more conscious in terms of content and delivery. Now new platforms facilitating virtual classes are emerging with features for effective learning and giving a satisfying experience to both teachers and learners. With some platforms also giving an equitable mechanism for evaluations, the potential of seeing a laptop as a full-fledged institute cannot be overlooked. Through a laptop, one can enroll aspirants, provide them instructions, evaluate them, and even finally issue certificates. There would be no exaggeration in extending the proposition to a time where individuals' tryst with education can match the Gurukul culture where learners have options to enroll with any cyber guru and adhere to his guidelines and live up to his evaluation process. The day may not be far off when in this consumeristic competitive world, stamped by a MOOC, will have a more satisfying professional life than those with certificates or degrees from 'some conventional universities'.

Clearly the stage is set for the emergence of an open, flexible, accessible, and inclusive system. The system will be governed on one hand by delivery-oriented teachers, and on the hand by keen learners and not mere enrolling students. The realization of virtuality will be the ultimate gain and satisfaction for teachers as well as learners.

## REFERENCES

Anand, Satyapal (1979). *University Without Walls*. Vikas Publishing House: New Delhi.

Bhatnagar, Satyavan (1997). *Distance Education: A System under Stress: An In-depth Study of the Indian Institute of Correspondence Courses*. Concept Publishing Company.

Brassard, M.L. (1982). 'Glossary', in: Learning at a Distance: A World perspective. Daniel et al. (Ed.), A Edmonton: ICCE.

GrowInstis, (1900). Retrieved on 9 July 2020 from https://odeweb.files.wordpress.com/2020/07/growins tis-1900-1.pdf

GrowInstis (2001). Retrieved on 9 July 2020 from https://odeweb.files. wordpress.com/2020/07/institutes-2001.pdf

Harper-Hermod, (1881–6). Retrieved on 9 July 2020 from https://odeweb.files.wordpress.com/2020/07/harper-hermod-1881-6-1.pdf

History of the University of London, (2020). Retrieved on 9 July 2020 from www.london.ac.uk/history.html.

Holmberg, Börje (1995). 'The Evolution of the Character and Practice of Distance Education', In: Open Learning.

International Correspondence Schools of Scranton, Pennsylvania – 1891 to the Present (2020). Retrieved on 9 July 2020 from http://digitalservices.scranton.edu/cdm/history/collection/ics/

Jegede, O. & Naidu, S. (1999). Retrieved from www.ascilite.org.au/ajet/e-jist/docs/vol3no1/ editorial.htm

Keegan, D. (1986). Interaction and communication (Chapter 6, pp. 89–107). In: Keegan, D., *The Foundations of Distance Education*. Kent, UK: Croom Helm.

Mahajan, Ravi K. & Mahajan, Kalpana K. (2019). *Open & Distance Education in the New Millennium: A Chronology in Making*. White Falcon Publishing.

Mahajan, Ravi K. (1993). *Computers: A New Teaching and Learning Environment'*. Open Paxis, ICDE, UK. Vol. I.

Mahajan, Ravi K. (2011). ARIMA Modeling on Google Search Results on Terminologies of Non-Traditional Modus Operandi of Education. EduTech – eJoutnal of Education & Technology.

Mahajan, Ravi K. (2012). 50 Years on, Distance Education Still a Struggle, *The Tribune*, September 18, 2012. Retrieved from www.tribuneindia.com/2012/20120918/edu.htm#2

Mahajan, Ravi K. (2020). COVID19: The Conference Killer! www. thelifestylejournalist.in/covid-19-the-conference-killer/

Moore, M. G. & Shattuck, K. (2001). *Glossary of Distance Education Terms*. The Pennsylvania State University.

OdeSayings (1938). Retrieved on 9 July 2020 from https://odeweb.files. wordpress.com/2020/07/ odesaying-1900-1.pdf

Pioneers (1800). Retrieved on 9 July 2020 from https://odeweb. files.wordpress.com/2020/07/pioneers-1800. pdf

Radio (1920). Retrieved on 9 July 2020 from https://odeweb.files. wordpress.com/2020/07/radio1920s.pdf

TechContributors (1800). Retrieved on 9 July 2020 from https://odeweb. files.wordpress.com/2020/07/ techcontributors-1800.pdf

Terms (2000). Retrieved from https://odeweb.files.wordpress.com/2020/07/terms-2000.pdf

Thinkers (1900). Retrieved on 9 July 2020 from https://odeweb. files.wordpress.com/2020/07/thinkers1900. pdf

# 4 Covid-19 Pandemic and Paradigms of Education in Indian HEIs

## *A Mapping of Tectonic Shifts*

*Mohd. Rizwan Khan*

Department of English, Aligarh Muslim University, Aligarh,
Uttar Pradesh, India

## CONTENTS

> "…Knowing how way leads on to way,
> I doubted if I should ever come back."
>
> (Frost Web 1)

## 4.1  INTRODUCTION

Humans witness tectonic shifts in the world order whenever struck by a catastrophe. Each major event of human history changes the course of civilizational paradigms. History is a witness to the truism that some important changes and innovations in human lives have generally been adopted as temporary survival techniques in such times. But the message at the beginning of the chapter illustrates the fact that most of the changes cannot be reverted as one road leads to another. Contemporary human systems are going through some sudden changes adopted as a means to sustain in the Covid-19 pandemic. One of these systems is education. The present chapter is a mapping of the shifts witnessed in the system of education with a focus on India. The chapter is aimed at assessing and taking stock of the situation to facilitate (in a modest way) a worldview for policy and planning for the future.

## 4.2 BACKGROUND

### 4.2.1 THE PAIN AND THE IMPACT

The twenty-first century has witnessed its biggest tragedy yet in the face of the Covid-19 pandemic. Millions of people have lost their lives and there seems to be no end in sight since the pandemic keeps coming back with new waves of more lethal virus variants. So far, most part of the world has witnessed two waves and a third wave is inevitable in the months to come. Wrecked with the second wave of the pandemic, India is going through one of the most painful experiences second only to the horrendous partition of the subcontinent. Fear plagues people like never before because an invisible enemy is felt to be pervading the existence of human systems. The institution of medicine stands challenged to its core. Helplessness is writ large all over the underdeveloped world since the production and procurement of vaccines is in the control of a few developed nations. Every human institution stands impacted. Hopelessness and despair have become the most potential tormentors for those who witness their loved ones being lost in the painful darkness of untimely death caused by Covid-19. The virus has impacted survivors too in physiological as well as psychological ways.

### 4.2.2 WHY TALK OF THE INSTITUTION OF EDUCATION?

Since survival has become a priority for millions who have been pushed into poverty, despair, and pain, it may seem inconsequential to note the struggles of other civilizational institutions. But with the reality of the pain of millions, equally real is the human instinct to construct a dignified life against all odds. Therefore, it is important to talk and discuss other institutions along with medical science, since it confirms a commitment of humanity to the future and nurtures our lifeforce to survive and see tomorrow. Even though the impacts of the pandemic are innumerable, we cannot give up. Time will tell what impacts will last forever, but it is imperative on the part of those who are capable in the present, to discuss, talk, brainstorm, document, and envision for the future to usher in hope for humanity. This has happened in the past as well. During the Spanish flu universities and colleges underwent various challenges and emerged stronger and well equipped with knowledge, techniques, and tools to fight other similar events like the polio outbreak, SARS, Ebola, etc.

### 4.2.3 NEW REALITY: ONLINE MODE

Life has undoubtedly changed in the Covid pandemic. With the discourse of "social distancing," "online mode," and "work from home" (WFH), workplaces have undergone monumental change. For more than a year now, most professional establishments have been running with the WFH model. Things and actions that were impractical only a year back have now become mundane reality. The institution of education too has witnessed tectonic shifts in the course of teaching, learning, and evaluating. Online mode emerged as an agency of sustenance for academic systems world over during the pandemic.

### 4.2.4   How IT Began

In India, online teaching is not an altogether new system since various virtual teaching and learning platforms have been present for some years now. For example, online private coaching institutes, online learning platforms, and skill-enhancement online courses disseminated through BYJU, Unacademy, Coursera, Vedantu, and several others have been present in the country for many years now. The Internet and its innumerable apps have also been known to Indians as in the rest of whole world. Yet, there has always been apprehension towards the online mode of education. Online education has merely existed as a subordinate and additional option to facilitate education for those who wish to go an extra step in their academic pursuits. Online pedagogy was never expected to become mainstream as far as the Indian scenario of education is concerned. While the government had been trying to create awareness and introduce various online learning platforms before the pandemic, it yielded a very low rate of acceptance among universities and colleges. However, it took the Covid pandemic only a few weeks to transform the apprehension of the stakeholders into acceptance of the online education model as a saviour in times of crisis. To quote Rizwan Khan in this regard:

> Over the last couple of years, the ministry of human resource development, the University Grants Commission and various other regulatory agencies have been attempting to ensure quality in education with the help of accreditation and ranking agencies. But these efforts always met with systemic resistance. Today, because of Covid-19, online pedagogy, among other mantras of quality in higher education, has come to the fore. Many options like massive open online courses, credit transfer and e-governance have remained under utilised by institutions. But the pandemic has left us with no option.
>
> Web 3

Thus, stakeholders' hesitation towards online mode has been replaced by a willingness to embrace it. Khan observes, "There are visible changes in the attitudes of teachers, students and other stakeholders to create a mental and physical viability for online pedagogy" (Web 3). Hence, it is now time for taking stock of the situation to augment online education in the country because "Now that we have realised that online teaching is going to stay on as the future mode of education, government agencies and institutions need to develop policies, provisions and practices for it" (ibid Web 3).

### 4.3   WHAT HAS HAPPENED DURING THE PANDEMIC SO FAR?

In March 2020, the pandemic knocked at the doors of the country and to curb its onslaught a lockdown was announced by the government leading to the shut-down of colleges and universities all over India. The University Grants Commission (UGC) and the Ministry of Human Resource and Development (MHRD) came into action and started circulating guidelines and regulations to shift to alternative modes of conducting the teaching/learning. The guidelines were given on March 5, 14, and 19, 2020 in its letter to the universities by the UGC. It states, "While it is crucial to follow measures taken by the Government to contain the spread of COVID-19, it is also important to continue the educational process making effective use of technology and other available options" (Web 9, 2). In the initial stage of the lockdown period a good number of teachers began to use apps like WhatsApp in order to communicate with students and upload their lectures along with the study materials. The "UGC Guidelines on Examinations and Academic Calendar for the Universities in View of Covid-19 Pandemic and Subsequent Lockdown" discusses among various modalities the Modes of Teaching-learning Process and the Modes of Examinations. One of the important observations in the guidelines is:

The faculty members have contributed a lot for the benefit of students during the lockdown period by using a number of tools like WhatsApp groups, other social media tools and emails. But the students also expect that the faculty must maintain a "substantive contact" with them. So, even after posting the lecture material online, teachers need to maintain communication with the students and discuss course material with them on a regular basis (Web 9, 3).

It is important to note that online teaching or e-learning are promptly recommended for the sustenance of academic procedures since only uploading lectures on social media apps is insufficient. To quote a few important extracts as case in point:

The MHRD and the UGC have been emphasizing to continue with the teaching-learning process using online modes such as Google Classroom, Google Hangout, Cisco Webex Meeting, YouTube Streaming, OERs, SWAYAM Platform, and SWAYAMPRABHA (available on Doordarshan (Free dish) and Dish TV), etc. (Web 9, 3).

The universities may conduct the Ph.D. and M.Phil. Viva-Voce Examinations through Video Conferencing using Google, Skype, Microsoft Technologies or any other reliable and mutually convenient technology, subject to the approval of the concerned statutory authority of the university, in compliance of Clause 9.6 to 9.9 of the UGC Regulations, 2016 regarding award of M.Phil/Ph.D. Degree. While conducting the Viva-Voce Examination through Video Conferencing ... (Web 9, 7).

It is evident from these extracts that despite the exigency, there was no dearth of options for teachers to choose from in place of offline mode of teaching. The guidelines also discuss relaxation with regard to attendance, evaluation, and modalities of e-lab-work among various other issues. Since the commencement of the pandemic lockdown, various such communiqués have been received in the universities and colleges regarding teaching and learning. A list of sample communiqués is provided at the end of the chapter.

### 4.3.1  ONLINE TEACHING/LEARNING AND E-LEARNING BEFORE THE PANDEMIC

In India, online teaching/learning or e-learning has been very bleak notwithstanding the fact that the ICT changes and developments, innovations in the pedagogy have been happening for over two decades now. Ranging from excellent to average, most of the academic institutions have had some exposure to ICT as well as internet-oriented infrastructure. The possibility of online education or e-learning would have been bleak if the pandemic had struck two decades back. Despite several efforts on the part of the regulatory bodies' engagement reliance on online technology has remained minimal if a bird's eye view of the country is taken. The different platforms that already existed and yet remained much under-utilised are as follows:

- SWAYAM Online Course (http://storage.googleapis.com/unique courses/online.html)
- UG/PG MOOCs (https://ugcmoocs.inflibnet.ac.inugcmoocs/moocs _courses.php)
- e-PG Pathshala (https://epgp.inflibnet.ac.in)
- e-Content Courseware in UG Subjects (http://cec.nic.in/cec/)
- SWAYAMPRABHA (www.swayamprabha.gov.in)
- CEC-UGC YouTube Channel (www.youtube.com/user/cecedusat)
- National Digital Library (ndl.iitkgp.ac.in)
- Shodhganga (https://shodhganga.inflibnet.ac.in)
- e-Shodh Sindhu (https://ess.inflibnet.ac.in)
- Vidwan (https://vidwan.inflibnet.ac.in) (Web 4)

Some of the other resources that already existed:

- Google Meeting
- YouTube
- Hangout
- Cisco Webex

## 4.4 TEACHING

### 4.4.1 CHALLENGES

Like most professionals, teachers also have had one of the most radical experiences of their professional lives during the pandemic. In terms of handling online teaching, their experiences range from extremely hostile to extremely gratifying. It began with a catastrophic announcement, the world over as well as in India, of lockdown. On the one hand, teachers struggled with fear like all other people and on the other hand they were looked up to as saviours who would be able to magically spin out some ways to satisfy students and their families' expectations. People of the world today including teachers of the world today, barring a few centurions, have never experienced a pandemic before. Thus, for teachers it was the directives of the regulatory bodies along with their own sense of responsibility that resulted in them initiating online teaching. But within no time the teachers found their space in the technology through trial and error. It was the YouTube videos and other resources on social media that came to the rescue of teachers and allowed them to learn new apps and platforms. Many teachers hoped for a uniform online learning app or platform provided by their institution, but in most cases had to decide independently what app or platform to use. The inability to physically visit the workplace and interact with colleagues also made the transition to online learning more difficult.

### 4.4.2 SHIFTS

This scenario has given birth not only to various shifts but also to possibilities of further desired shifts in the future. Teaching in person has been a central tenet of the education system in India. Adoption and execution of online mode of teaching have made it possible to shift from this ethos. Teachers of different age brackets have taken their own time to come to terms with it. But having gone through the challenges, most teachers are much better informed about various platforms and apps for online teaching as compared to before. Teachers have also found ways to optimize their teaching through the use of technological aids instead of just relying only on books and offline sources. This has opened vistas in education and made education more impactful. The shift to online official meetings as well as interaction with colleagues also allowed teachers to connect with their workplace. Thus, online learning has empowered the paradigm of autonomy for teachers.

## 4.5 LEARNING

### 4.5.1 CHALLENGES

Like teachers, students too faced a dilemma at the commencement of lockdown. On the one hand they wanted to keep up with their studies, but on the other they did not know what was going to happen to their classes and how the teaching would continue, leaving them confused and concerned about their futures. While teachers tried their best to share notes, materials, and lectures through social media apps like WhatsApp, in the beginning students worried about the quality of their education. Fortunately, online teaching learning platforms were started in many institutions and provided the continuity students needed. One of the challenges, however, has been the fact that not all students have accessed to laptops, smartphones, or even the internet. If there are several students in a family more hardware is needed in addition to more internet data. This made it hard for some students to continue their education on a regular basis as they had previously. Many lower-middle and lower-income families in India cannot afford to buy multiples devices or more bandwidth. Moreover, some families lost income during the pandemic, making the situation even more challenging. In addition to this, many regions of the country do not have access to high-speed internet at all, making online learning impossible. Students also missed the company of their peers and their campus

### 4.5.2 Shifts

As has been seen, learning has been transformed into an entirely different experience because of the shift to online learning. But this shift has been mostly positive. To be able to make presentations on a regular basis using various apps like Google Meeting, Zoom Meeting, Blackboard, Webex, etc., in the process of learning generates a sense of involvement among learners. Online learning also allows and encourages consistent engagement with online tools of learning like consulting multiple web resources while preparing assignments and presentations and collaborative/teamwork with other students. Students are also empowered because they have adapted to life in the midst of a major catastrophe while handling their learning and studies.

## 4.6 EVALUATING

### 4.6.1 Challenges

When the pandemic and the subsequent lockdown struck, the academic session was at its end in most of the terminal semesters across universities in India. Doubts and uncertainty hovered over all the students and especially the terminal semester ones about the question of their degrees. When, how, and what type of evaluation would be done so students could attain their degrees and progress in their programmes were among the questions that had to be answered. Students did not have training and experience taking their tests through alternative modes like online tests, therefore, they displayed a lot of apprehension. Moreover, over-leniency in grading and marking occurred, which may prove detrimental in the coming years.

### 4.6.2 Shifts

But it is in times like these that innovations come to the fore thick and fast and "normal" is replaced. Various formative assessments like MCQ/OMR-based examinations, Open Book Examination, Open Choices, and assignment/presentation-based assessments stakeholders had been waiting became functional during the pandemic. This is visible in the following extract from the UGC guidelines:

1. Maintaining the sanctity of academic expectations and integrity of examination process, the universities may adopt alternative and simplified modes and methods of examinations to complete the process in shorter period of time in compliance with CBCS requirements as prescribed by UGC from time to time. These may include MCQ/OMR based examinations, Open Book Examination, Open Choices, assignment/presentation-based assessments, etc.
2. The universities may adopt efficient and innovative modes of examinations by reducing the time from 3 hours to 2 hours assigned to each examination, if need arises but without compromising the quality, so that the process may be completed in multiple shifts and, at the same time, sanctity to evaluate the performance of a student is also maintained.
3. The universities may conduct Terminal/Intermediate Semester/Year examinations in offline/online mode, as per their Ordinances/Rules and Regulations, Scheme of Examinations, observing the guidelines of "social distancing" and keeping in view the support system available with them and ensuring fair opportunity to all students.
4. Terminal semester/year examinations for PG/UG courses/programmes may be conducted by universities as suggested in the academic calendar keeping in mind the protocols of "social distancing."
5. For intermediate semester/year students, the universities may conduct examinations, after making a comprehensive assessment of their level of preparedness, residential status of the students, status of COVID-19 pandemic spread in different region/state and other factors.

In case the situation does not appear to be normal in view of COVID-19, in order to maintain "social distancing," safety and health of the students, grading of the students could be composite of 50% marks on the basis of the pattern of internal evaluation adopted by the universities and the remaining 50% marks can be awarded on the basis of performance in previous semester only (if available). The internal evaluation can be continuous evaluation, prelims, mid-semester, internal assignments, or whatever name is given for student progression. In the situations where previous semester or previous year marks are not available, particularly in the first year of annual pattern of examinations, 100% evaluation may be done on the basis of internal evaluation. If the student wishes to improve the grades, he/she may appear in special exams for such subjects during next semester. This provision for intermediate semester examinations is only for the current academic session (2019–20) in view of COVID-19, while maintaining safety and health of all the stakeholders and sanctity and quality of examinations (Web 9, 5–6).

The above extract also indicates the significance of flexibility in the system of examinations across universities in the country. The major shift that one must point out is best expressed in Khan's words: A significant paradigm is the examination system. At present, most institutions are perplexed with regard to conducting examinations. The pandemic has given us an opportunity to revisit the entire idea of examination, testing and evaluation. We need to decide whether we want to test or assess our students (Web 3).

Pedagogues need to consider that evaluation not only has a washback effect but also leads to the fair opportunity of assessment of different types of learners as well as of different disciplines.

## 4.7 WHERE DO WE GO FROM HERE?

In April 2020, in its report and recommendations for handling the higher education institutions (HEIs), UNESCO recommended the following:

1. Anticipate a long-term cessation, focusing efforts on ensuring teaching continuity and guaranteeing equity, generating governance mechanisms, monitoring, and efficient support;
2. Design pedagogical measures to evaluate training and generate mechanisms to support learning for disadvantaged students;
3. Document the pedagogical changes introduced and their impacts;
4. Learn from mistakes and scale up digitization, hybridization, and ubiquitous learning; and
5. Promote internal reflection on the renewal of the teaching and learning model (UNESCO, 2020).

In the light of these recommendations and the aforementioned discussion of teaching, learning, and evaluation we know various kinds of shifts are imminent. Although many of these were contingent shifts many changes will be permanent as one road leads to another. It is important to look ahead to plan for updated systems of teaching, learning, and evaluating. The HEIs in India are going to be very different in the coming years from what they look like today. In the words of Rizwan Khan, "Now that we have realised that online teaching is going to stay on as the future mode of education, government agencies and institutions need to develop policies, provisions and practices for it" (Web 3). Let us take a stock of the tectonic shifts and how these are going to impact the education system in the country.

- In the traditional systems, teaching and learning are embodiments of lecturing and listening, respectively. The online/e-pedagogy requires teaching and learning to be collaborative exercises as teachers and students go beyond their traditional roles to participate actively in the exercises, activities, and instructional materials uploaded online.
- We need to look at the present moment as an opportunity to optimise the weeding out of the outdated methods from the pedagogical systems. We often find century-old colleges and

universities in dilapidated conditions due to insufficient funds, poor teacher student ratio, and apathy of the system. The education budget of the country has not increased to match the expanding needs of the education sector. Online teaching can be used in a cost-effective way to reach out to those students who do not have access to teachers in their colleges/universities. Such students can access online the teachers who are available in different parts of the country. We have waited long enough for a tomorrow where we shall have sufficient funds to transform each college and university into equipped organisations with the help of funds. This wait has made generations of youth lose access to quality education. A much cheaper way to resolve this mammoth problem is to engage with and reach out to all the students through online mode of learning and ensure quality education for them

- Often we find students complaining about the lack of infrastructure as a reason for their underperformance or poor performance. Virtual pedagogy can address this issue too. In a survey based article on an Australian university about the H1N1 pandemic similar concerns and shifts have been observed among students:

Online resources such as lecture recordings and forum tutorials allow for off-campus education, and can provide continuity of learning for students undergoing isolation. However in our study, few respondents had adopted the use of online teaching or learning resources as a result of pandemic influenza (H1N1). This may be due to a number of factors including: (1) the apparent mildness of the pandemic; and/or (2) the lack of promotion by the University to use these resources. It was encouraging to see that students were very willing to continue University schooling via online resources, indicating the potential for expanding the existing UNSW online teaching resources. While it was encouraging that students would undertake online courses, we found very little support for an online teaching method among the academic staff members. Reluctance to use online resources was associated with increased age, and may be due to unfamiliarity with or resistance to technology. In preparation for an outbreak, universities should focus on creating additional support for technologies that allow faculty and students to continue their teaching and learning activities that minimise disruption. Online recordings, virtual learning environment, blogs, web conferencing, and discussion forums should all be utilised to assist in the delivery of lessons. Having a contingency and communication plan for teaching key sections may provide the needed continuity for students and faculty. Training must be provided in the pre-pandemic periods to minimise disruption.

- Use of television networks of the public broadcasters can be a tool to optimise the outreach of education. The prerecorded lab manuals, lectures, experiments, and demonstrations can be telecast in a more effective and programmed manner to facilitate the learning of the students who cannot access the internet and gadgets. In fact, such recording can be saved in the social media networks like YouTube where students can access them free whenever they need to access.
- It is very crucial to understand that no singular system should be relied upon as the absolute paradigm to educate. A combination of online, offline, live, recorded, and televised access to learning is an ideal way to maximise the accomplishment of learning outcomes.
- Access to gadgets, the internet, and tools to engage in online teaching and learning for teachers and students is a big elephant in the room and we need to address it. If we assess the reduction in running expenses of the universities and colleges that has happened during the lockdown period, we will surely find that a lot of money can be saved and in future this money can be channelized to ensure access of gadgets and internet to those who cannot afford it.
- The question of the non-availability of the internet also needs to be handled in an innovative way. Instead of procuring physical space for college/university, future planning in education needs to invest in procuring virtual space for the expansion of the universities to reach in every

nook and corner of the country's landscape lest no part remains on the blind-spot of the virtual world. Through diversion of funds into the procurement of virtual space academic and educational equity can be cemented.

• Universities and colleges need to move towards a model of dual mode wherein online and face-to-face teaching/learning/evaluating shall coexist. The possibility of this model has become a reality in many institutions across the country and the regulatory bodies are also set to encourage the same. To quote from a UGC communique:

In order to overcome such challenges in the future, the faculty should be adequately trained for the use of ICT and online teaching tools, so that they complete about 25% of the syllabus through online teaching and 75% syllabus through face-to-face teaching.

Web 9, 10

The pandemic has shown us that it is never too difficult to turn the tables around when it comes to making the evaluation an innovative paradigm. To quote Khan:

Testing and evaluation need to be subordinated with the spirit to assess a student rather than to pass a judgment on a student. Assessment is more constructive and positive. We must not forget that the examination system has a washback effect on pedagogy. When our students are tested, they study to pass the test, but when our students are assessed they study to perform. Therefore, we must switch over to assessments that are formative. The pandemic is likely to have a lasting impact on our lives and minds. Therefore, we need to minimise the burden of this impact by creating an ecology where judgmental and penalising tendencies are weeded out.

Web 3

Thus, following aspects of assessment and evaluation emerge that define and underline the existing practices and possible ways to improve the same in the Indian HEIs:

• In the pursuit of quality assurance in academics, assessment is the keyword because testing becomes a strait-jacketed hegemonic institution of classifying students as good or bad on the basis of a few hours of the pen-paper test. In fact, it is life that tests learning and education. An education institution can only assess the accomplishment of learning outcomes among students so that the teaching can be revisited for improvement. Multiple assessment techniques spread over the whole semester lead to equitable opportunities for students to excel and perform in their area of expertise and comfort.

• Formative assessment practice impacts the critical thinking of students and teachers. Critical thinking is a crucial discourse in contemporary education systems and needs to be taken into cognisance during the policy and planning of academic charters. A teacher undergoes exposure to assess the students through innovative assessment techniques that the formative evaluation is comprised of. Such techniques emphasise on assessing students' growth at six levels defined in Bloom's Taxonomy as can be seen in Figure 4.1.

• The growth of a student from the "remember" level to the "create" level involves dozens of actions that are to be ensured in the process of teaching and learning. Only assessment of formative nature can facilitate the attainment of this action-based growth of students and teachers as critical thinkers.

• Another major concern in this situation pertains to the students who have dropped out of colleges and universities due to inaccessibility to online learning, familial circumstances like losing family members to Covid-19, depression, poor financial conditions, and other reasons. Such students need to be brought back to the classes with more care and designated counselling portals created by the regulatory bodies. Although the regulatory bodies have displayed response towards grievance redressal of the students, this needs to be done more pro-actively

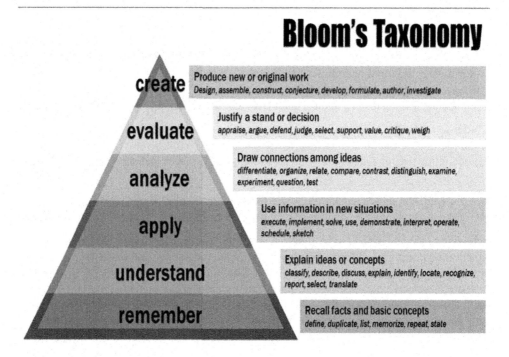

**FIGURE 4.1**   Bloom's taxonomy.

*Image Source*: Vanderbilt University Center for Teaching Web 1 open source.

because after the very lethal second wave the number of drop-outs and problem of absenteeism has become manifold now. If this issue is allowed to linger on we will lose these students forever from our classrooms. Interestingly, in a study of epidemics and impacts on HEIs the findings note that the connection of students with their institutions increased during and after epidemics:

Yet these historical records also contained hope – while a few institutions suffered dire losses due to the diseases of 1878 and 1918, the records also noted increased connection to institutions. As universities and colleges serve various research, social, and community needs, what will be the long-term impact of this new pandemic on higher education?

Foster and Thomas, 2020

It is yet to be ascertained in the present pandemic whether students and teachers have strengthened their connection with their academic institutions. But a need for a refurbished role of the college/university in the lives of stakeholders and society is already felt to engender a more educated and informed response to such situations rooted in preparedness.

- Research needs to be carried out as national projects among universities and colleges to assess and analyse how well and bad the Indian HEIs have fared in their response to the pandemic. A repository of interviews, surveys, and other relevant data can be created without spending much in order to make it accessible for future researchers. This kind of research will project an audit of the pandemic preparations that many universities have been participating for many decades in countries like the United States, Australia, Singapore, and in Europe.
  - Last but not least, the role of a university campus during such and other catastrophes has come to be realised as crucial. The university/college campuses are the places where the

discourses and generations of the future are shaped and constructed. We have seen in the past that during Spanish flu, world wars, and other major tragic events students and teachers as a community play a role in healing of society in physiological as well as psychological ways. Such a phenomenon has yet to emerge in the Indian context in a strong manner. The absence of access to campus has hit the academic community hard enough to make them forget that a campus is more than a physical entity. An academic campus extends beyond the buildings and infrastructure. We need to consolidate that campus beyond the buildings. The concept of online and virtual campus needs to be strengthened to build a more empowered system of education that does not take a hit by such calamities in the future.

## 4.8   CONCLUSION

The pain and trauma of the pandemic hit us hard and threw us into uncertainties, but we have come far enough that we can stand and look back to assess the situation to create a better present and to be well equipped for tomorrow. While history shows us that pandemics have happened and will happen in the future it has also shown us that human beings have also not stopped progressing, learning, and educating. Education is one of the pillars of this human journey of survival and sustenance. A reckless attitude towards education may cost us irretrievable losses. We need to reiterate that education is a paradigm of relentless revisiting to ensure qualitative learning and teaching. To quote Van et al.:

> Universities are not immune to natural or manmade disasters, and past experience with these have illustrated the importance of continuity during and after these events [5, 9]. In a ... pandemic, ... institutions must maintain a balance between academic continuity, with infection control and minimising morbidity [5] ... universities must invest in online teaching resources and training during inter-pandemic periods. There also needs to be greater recognition for the need for online assignment submission and examinations to ensure minimal disruption to the students.
>
> 2010

Pandemics have taught us that education should never be de-prioritised. A pandemic is also a reminder to countries likes ours that we need to prioritise qualitative education to be able to compete in the world. It is evident from the shifts that have taken place in the country's education scenario that it will be mandatory for us to continue with many of these changes. When such changes are visible in the system, monitoring becomes equally important to maintain quality because:

> When innovations occur, quality check has to follow as a response system to fructify the results of innovative pedagogy. Institutions and stakeholders need to regularly assess what an innovation facilitates. The answer to this question will lead to quality assurance. Regular stocktaking through a transparent feedback system is a handy tool to conduct such validity checks.
>
> Khan, Web 3

Negotiations among technologies, stakeholders, and regulatory bodies must be routine. As Tennyson says,

> The old order changeth, yielding place to new ...
> Lest one good custom should corrupt the world.
>
> Tennyson, Web 5

## REFERENCES

Armstrong, P. (2010). Bloom's Taxonomy. Vanderbilt University Center for Teaching. https://cft.vanderbilt.edu/guides-sub-pages/blooms-taxonomy/. URL: www.flickr.com/photos/vandycft/29428436431 (Accessed on May 17th, 2021)

Foster, H. & Thomas, J. (2020). Higher Education Institutions Respond to Epidemics. *History of Education Quarterly, 60*(2), 185–201. doi:10.1017/heq.2020.11 (Accessed on May 19th, 2021)

Frost, Robert. "Road Not Taken." www.poetryfoundation.org/poems/44272/the-road-not-taken (Accessed on May 15th 2021)

Khan, M. Rizwan. "Assess Students, Do Not Test Them," *The Week*, August 30, 2020, www.theweek.in/thew eek/cover/2020/08/20/assess-students-do-not-test-them.html (Accessed on May 16th, 2021)

"On-Line Learning-ICT Initiatives of MHRD and UGC." 25th April, 2020. URL: www.ugc.ac.in/pdfnews/ 1573010_On-Line-Learning---ICT-initiatives-of-MHRD-and-UGC.pdf (Accessed on May 17th, 2021)

Tennyson, Lord Alfred. "Morted'Arthur." URL: www.poetry foundation.org/poems/45370/morte-darthur#:~:text= %22The %20old %20order%20changeth%2C%20yielding,custom%20 should%20corrupt%20the%20 world. (Accessed on May 17th, 2021)

"UGC Guidelines on Examinations and Academic Calendar for the Universities in View of Covid-19 Pandemic and Subsequent Lockdown" 29th April, 2020. URL: www.ugc.ac.in/pdfnews/4276446_UGC-Guideli nes-on-Examinations-and-Academic-Calendar.pdf (Accessed on May 16th, 2021)

UNESCO: COVID-19 and Higher Education: Today and Tomorrow: Impact Analysis, Policy Responses and Recommendations (April 9, 2020), UNESCO. www.iesalc.unesco.org/en/wp-content/uploads/2020/04/ COVID-19-EN-090420-2.pdf (Accessed on May 20th, 2021)

Van, D., McLaws, M.L., Crimmins, J. *et al.* (2010). University life and pandemic influenza: Attitudes and intended behaviour of staff and students towards pandemic (H1N1) 2009. *BMC Public Health* **10,** 130. https://doi.org/10.1186/1471-2458-10-130 (Accessed on May 19th, 2021)

## WEB RESOURCES

www.ugc.ac.in/pdfnews/6838376_9964222_HEI-recognition-2019-as-30-01-2020.pdf

www.ugc.ac.in/pdfnews/6270462_MOOCs.pdf

www.ugc.ac.in/pdfnews/7248697_UGC-Advisory---Permission-to-work-from-home-in-view-of-COVID-19.pdf

www.ugc.ac.in/pdfnews/4276446_UGC-Guidelines-on-Examinations-and-Academic-Calendar.pdf

www.ugc.ac.in/pdfnews/5369929_Letter-regarding-UGC-Guidelines-on-Examinations-and-Academic-Calen dar.pdf

www.ugc.ac.in/pdfnews/4935932_letter-SOP-8th-July.pdf (July 8th, 2020)

www.ugc.ac.in/pdfnews/5660666_ANNEXURE-I.pdf (July 8th, 2020)

www.ugc.ac.in/pdfnews/6695616_ANNEXURE-II.pdf (July 8th, 2020)

https://www.ugc.ac.in/pdfnews/9514949_English.pdf (July 18th, 2020)

https://www.ugc.ac.in/pdfnews/7438482_LETTER--Provisional-admissions-and-submission-of-documents-of-qualifying-examination.pdf (August 31st, 2020)

https://www.ugc.ac.in/pdfnews/9919836_offline-exams-letter.pdf (May 6th, 2021)

# 5 Shifts in Education and Career Choice of Students in the Post-COVID-19 Era

*Rupinder Bir Kaur\* and Kanwal Roop Kaur*
University Business School, Panjab University, Chandigarh, India
\*rupinderbir@gmail.com

## CONTENTS

## 5.1   INTRODUCTION

COVID-19 crisis sheds light on the need for a new education model.

*- Education Times*, April 13, 2020

Based on the severity of the COVID-19 pandemic and existing health framework, many countries have adopted different measures regarding the spread of the virus starting from mild restrictions to extreme lockdown. The COVID-19 pandemic has affected the lives of people physically, economically, and psychologically. In this chapter, our focus is on the impact of this outbreak on students' attitude and perception towards their education. All the universities, colleges, schools, and educational institutes have been on lockdown as they can become outbreak centres very easily. Therefore, the trend towards e-learning and virtual classes is increasing. Harvard and many other universities have opened up a number of online courses for free enrollment. The concept of e-learning has indeed created opportunities as well as challenges for both students as well as teachers.

### 5.1.1   DEFINITIONS OF THE TERMS USED

Families: "Families are central to education and are widely agreed to provide major inputs into a child's learning," as described by Bjorklund and Salvanes (2011).

Pandemic: A pandemic is defined as "an epidemic occurring worldwide, or over a very wide area, crossing international boundaries and usually affecting a large number of people."

Attitude: "Attitude can be defined as how a person views and evaluates something or someone, a predisposition or a tendency to respond positively or negatively toward a certain idea, object, person, or situation."

DOI: 10.1201/9781003358916-5

Perception: "Perception is a mode of apprehending reality and experience through the senses, thus enabling discernment of figure, form, language, behavior, and action."

### 5.1.2  MAIN QUESTIONS

This work aims to address the following questions:

- Amidst the outbreak, what are students planning to do?
- How has the outbreak affected the career aspirations of students?
- What changes are there in student attitudes after this outbreak?

## 5.2  PARADIGM SHIFT

### 5.2.1  FROM OFFLINE TO ONLINE

The shift should be towards expanding online teaching and learning resources for sustaining education amid disaster. To attract the attention of students to use different and more scientific-based information sources, new policies must be framed. Also, efforts should be made to boost the e-teaching infrastructure, especially in the low-income private and government educational institutions which had to be closed (due to lack of proper infrastructure), posing a great challenge for institutions, teachers, and students.

According to UNESCO's COVID-19 Educational Disruption and Response, "Most governments around the world have closed educational institutions in an attempt to contain the spread of the COVID-19 pandemic. These nationwide closures are impacting almost 70% of the world's student population." The shift from offline to online has not only brought a transformation in the education system but has also raised new opportunities for small entrepreneurs to move online (www.unesco.org/en/articles/covid-19-educational-disruption-and-response).

### 5.2.2  THE PSYCHOLOGICAL IMPACT

It was reported that individuals who spend their time watching movies and on social media have higher psychological distress than those who have someone to talk with (Arënliu and Bërxulli, 2020). Therefore, during epidemics the mental health of college students should be monitored effectively (Wenjun Cao et al., 2020).

UNESCO has made some recommendations including "providing support to teachers and parents on the use of digital tools," "developing distance learning rules and monitor students' learning process," and many more.

### 5.2.3  GROWTH IN IT SECTOR

The paradigm shift from offline to online predicts the rise in computer technology and the information technology sector. An innovative approach and emphasis on this sector is the need of the hour. This sector has already attracted the interest of Indian youth and will see further attention.

### 5.2.4  AGRICULTURE AND FARMING

The pandemic and its resulting lockdowns have demanded demanded automony of all countries of the world. Being self-reliant in a country like India means identifying the core competency of our country: agriculture and farming. The exploitation and sustainable care of natural resources and focus on agriculture and farming needs to be encouraged among Indian young students. The stereotypes of linking illiteracy with agriculture need to be broken and more and more educated youth should be encouraged to enter into the realm of social and agricultural entrepreneurship.

## 5.3 CONCLUSION

To conclude, it has become highly necessary to determine paradigm shifts in different sectors of business, education, and career choices among students in India resulting from the pandemic and its consequences such as lockdowns. The transition period and post-COVID-19 period have resulted in the emergence of a "new normal." This new normal requires a focused change in education and careers of students.

## REFERENCES

Akan, H., Gurol, Y., Izbirak, G., Ozdatl, S., Yilmaz, G., Vitrinel, A., & Hayran, O. (2010). Knowledge and attitudes of university students toward pandemic influenza: A cross-sectional study from Turkey. *BMC Public Health*, 10, 1–8. http://doi.org/ 10.1186/1471-2458-10-413

Arenliu, A. & Bërxulli, D. (2020). Rapid assessment: Psychological distress among students in Kosovo during the COVID-19 pandemic.

Bjorklund, A. & Salvanes, K. (2011). Education and family background: mechanisms and policies. *Handbook of the Economics of Education*, 3(2), 34–39.

Cao, W., Fang, Z., Hou, G., Han, M., Xu, X., & Dong, J. (2020). The psychological impact of the COVID-19 epidemic on college students in China. *Psychiatry Research*, 287 (March), 112934. https://doi.org/ 10.1016/j.psychres.2020.112934

Van, D., McLaws, M. L., Crimmins, J., MacIntyre, C. R., & Seale, H. (2010). University life and pandemic influenza: Attitudes and intended behaviour of staff and students towards pandemic (H1N1) 2009. *BMC Public Health*, 10. https://doi.org/10. 1186/1471-2458-10-130

# 6 An Impact Assessment of COVID-19 Pandemic on the Higher Education Institutions of India

*Santosh Kumar\* and Abhishek Pandey*

Department of Geology, Centre of Advanced Study, Kumaun University, Nainital, Uttarakhand, India
\*skyadavan@yahoo.com

## CONTENTS

## 6.1   INTRODUCTION

Education is an essential element without which an individual, society, and country cannot grow and prosper. Education systems make citizens, societies, organizations, and nations locally and globally sustainable and competitive. Higher education institutions (HEIs) are open system spaces for universal tertiary education, and provide learner-centric environments to nurture and cater to education in particular disciplines or domains imparted by experts or specialists and advisors in the respective subject-fields. The HEIs, particularly the residential ones, provide an environment of continuous learning and interactions between the students and teachers in classrooms and in various forums such as workshops, seminar, symposia, high-end training, and internship programmes. Research students and scientists formulate hypotheses based on existing knowledge gaps and test them by carrying out experiments, ultimately contributing to the social and economic development of a nation (Arora, 2016; Kumar, 2016). Such learning processes and experimentation in the HEIs in India disrupted and suffered measurably by the COVID-19 pandemic (e.g., Edmund, 2020). In this chapter, the impact of the COVID-19 pandemic on teaching-learning-evaluation in HEIs and future career opportunities and placements of students are discussed along with suggestions.

## 6.2   PANDEMICS IN THE PAST

Pandemics such as Black Death or the Plague and Smallpox have affected countries throughout the world in the past including India, which was among the worst affected nations. Pandemics adversely affect social, cultural, and economic institutions. HEIs in particular are affected in significant ways (Adam, 2020; Gettleman and Schultz, 2020). In general, pandemics bring panic and instability that often drive institutions to take counter-productive measures. In past pandemics, faculty and students fled from campus for safety reason and could not return back to campus until the disappearance of pandemics. This resulted in substantial drops in enrollment. However, today we live in an entirely different era, and technology allows distance learning and other educational opportunities not available before.

## 6.3   DECLINE IN STUDENT ENROLLMENT AND REDUCED EMPLOYMENT OPPORTUNITIES

In India there are more than 1000 private and government-owned conventional universities, deemed universities and institutions of national importance, and nearly 45,000 degree colleges affiliated with universities for over 1.39 billion people (as per estimation in 2021). As per the information recently released from the Ministry of Human Resource Development (MHRD), now the Ministry of Education (MoE), Government of India, India's student population reached 37.4 million in 2018–2019. Gross Enrollment Ratio (GER), which refers to the percentage of students in higher education of the total eligible population in the 18–23 age group, has been set to increase from 24.5 (2015–2016) to 32 by 2022. The pandemic has adversely affected the targets or goals set by the students, teachers, and researchers in HEIs. The majority of 10+2 students of state and central boards, common feeder mass for the enrolment in the HEIs of India, has immediately faced problem during final year examinations which were postponed due to COVID-19 pandemic, and was rescheduled in July 2020, because of which the result were delayed at least by three months. Subsequently, a number of competitive examinations (JEE, NEET, CET, CLET, etc.) were also rescheduled or cancelled. Although most of the academic bodies like UGC, AICTE, BCI, PCI, NMC, etc., who act as nodal agencies for coordinating admission, teaching and examination, etc. between the central/ state governments and HEIs, have already published revised academic calendars, a large number of students are reluctant to join the courses because of days missed in previous semesters or due to fears of the spread of the virus. The Ministry of Education, Government of India has encouraged conducting the online/virtual classes. The net effect of the state of such uncertainties has resulted in a moderate to sharp decline in student enrollment in HEIs. Other factors related to economic conditions have contributed to drops in enrollment as well. On the other hand, in state- and central government-funded HEIs professional and conventional undergraduate courses have attracted the attention of a large number of students because of fee waivers or low and affordable tuition.

While increasing COVID positive cases in the second pandemic wave added to existing uncertainties, vaccination programs have improved the situation.

## 6.4   IMPACT ON INDIAN STUDENTS STUDYING IN HEIs OF INDIA AND ABROAD

While many Indian students are enrolled in various educational programmes in HEIs located in the United States, Canada, European countries, Japan, Australia, and south-east Asian countries. A small number of such students are provided fellowships or their studies are sponsored from India and host countries under academic exchange programmes, most students use their own resources or education loans from Indian banks. The loans to students are generally sanctioned on the basis of income of parents and/or mortgage of personal property. These students were evacuated from overseas under Vande Bharat mission scheme of Govt. of India in various phases by national aircraft carrier AIR INDIA that caused additional financial burden to students and parents. Although some concrete policies for easing payments of loan installment have been framed to minimize the financial losses, problems still persist. Doctoral and post-doctoral research students have been forced to shut down laboratory investigations and ongoing experimental works, causing huge financial and data loss. However, researchers involved in developing anti-COVID-19 vaccines were allowed to work continuously to cope with the pandemic.

## 6.5   POSTPONEMENT OF ENTRANCE, REGULAR SEMESTER, AND COMPETITIVE EXAMINATIONS

Some of the 10+2 passed students in India could not take entrance examinations in the year 2020, which are commonly held each year for admitting students in undergraduate courses of arts, social sciences, science, law, management, medical and engineering courses, etc. Most of the final year/semester examinations in HEIs are commonly held in the month of May/June each year. Due to COVID-19 there were delays of three to six months in the academic calendar of 2020–2021 sessions. The second wave of COVID-19 held back the academic session of 2021–2022, although the parallel vaccination program improved the situation. Delayed competitive entrance, state, and union public service examinations consequently limited employment opportunities.

## 6.6   ICT GROWTH, ITS RELEVANCE AND CHALLENGES DURING LOCKDOWN

The University Grants Commission (UGC) started countrywide digital classroom programmes in 1984, and for the production of educational materials, *educational multimedia research centres* (EMRCs) were set up in more than 22 universities across the country with major objectives to coordinate, guide, and facilitate educational programme production. The Information and Library Network (INFLIBNET) centre was established in 1991 as an autonomous Inter-University Centre of the UGC under MHRD, Govt. of India with the objectives of modernizing libraries and transferring and providing access to information, learning, and academic pursuits through a national network of libraries among the Indian Universities, Colleges, and R&D Institutions. Subsequently, the Consortium for Educational Communication (CEC) was established in 1993 as Inter-University Centres by the UGC with the goal of addressing the needs of higher education through the powerful medium of television combined with potential use of Information Communication Technology (ICT). The universities and colleges were continuously allocated huge funds until the end of 12th Five Year Plan (2012–2017) that is replaced and now supported under the *Rastriya Ucchatar Sikscha Abhiyan* (RUSA) for connecting campuses with broadband and fibre internet for easy and quick access to digital libraries and e-learning. In view of such infrastructure and to democratize the opportunities of quality education, the National Mission on Education through ICT (NMEICT) was launched to translate the power of IT into expanded learning opportunities that have the potential to change the higher education scenario. The use of all such ICT infrastructures and initiatives in the HEIs became imperative during the lockdown and postlockdown periods of the COVID-19 pandemic. As soon as COVID-19 emerged as a health crisis, the HEIs in India managed to shift quickly to online platforms and continue with classes. A number of opportunities are available to Indian students on various web platforms that can be availed at no cost that have high-speed internet facilities (Figure 6.1). For instance, a group of free to air 32 DTH channels education programme is fully devoted to telecasting on 24x7 basis using the GSAT-15 satellite.

The use of live lectures, educational videos, animations, quizzes, and project-based learning – Blended Learning – is the need of the present pandemic hours (Figure 6.2). Blended learning has been in discussions since 2001 (Reay, 2001; Morgan, 2002; Rooney, 2003; Takwale, 2003). Teachers equipped with basic internet connectivity can record their lectures and upload them on secure online platforms

**FIGURE 6.1** The audio–video modes of e-learning as innovative programs, which are operational in India. These can be used potentially by students to educate themselves. Apart from these, several other quality open education resources (OER) of overseas institutions (Stanford, Oxford, Cambridge, etc.) are also available to the students.

**FIGURE 6.2** Illustration of the concept of blended learning, which is derivative of traditional and ICT systems of learning.

preferably on their institutional website. These recorded lectures can be viewed offline at any time by students.

However, in India HEIs face a number of problems such as lack of widespread virtual classrooms for live lectures, lack of examination and evaluation procedures, poor internet connectivity in some remote localities and hilly regions, and lack of hardware such as laptops and smartphones. Other unintended outcomes of ICT-based teaching and learning include lack of supervision, outdoor games, field and laboratory works, educational trips, etc.

## 6.7   IS THE ICT A THREAT OR AN OPPORTUNITY TO HUMAN TEACHING?

Due to enhanced use of electronic digital media and online virtual classes during COVID-19, an academic discussion has emerged among users of ResearchGate and other forums through a number of webinars about whether the continuous use of ICT is a threat to traditional education. As noted by R. Bandara on ResearchGate teachers can use ICT to transfer knowledge and positively influence students and even perform in better ways using ICT by self-updating on subject matters, automating course material preparation, applying student-specific teaching methods, using automated assessments, and so on. But can ICT motivate, influence, and assess the creativity of students? These are the questions we do not yet have answers to. R. Bandara further explained that if ICT can completely replace a human teacher, then there will be no point in teaching the subject to humans as there will be intelligent machines that are experts in the subject they taught. O.A. Otekunrin argued on the same forum that ICT cannot replace teachers but can enhance teaching and learning processes.

## 6.8 IMPACT OF LOCKDOWN ON RESEARCH ACTIVITIES

While there have been uncertainties regarding what research activities are permitted to continue during the pandemic, the Ministry of Home Affairs (MHA) in October 2020 said that research scholars (PhD) and post-graduate students of science and technology streams requiring laboratory investigations can continue to work. However, the decision to open all the centrally funded HEIs is made by the Department of Higher Education in consultation with the MHA, based on the assessment of the prevailing situation at that time. Other HEIs such as state universities, private universities, etc., are under the jurisdiction of the respective state/UT governments. States/UTs usually prepare their own SOPs with regard to health and safety precautions.

Research activities were severely affected during the lockdown period of 2020, especially research in scientific domains requiring continuous monitoring, regular data collection, maintenance of controlled set of conditions (pressure, temperature, humidity), chemical testing, etc. Many problems were faced in life science research on cell culture, animal behavior monitoring, feeding, and cleaning of the animals in lab, and maintenance of humidity and temperature or light-dark cycles for studies of time-bound biological processes. The ongoing experimentations of chemistry labs were shut down completely and most had to be reinitiated. The research activities in earth sciences, which involve field work in distant locations, sample collection, analysis of samples in various laboratories located in India and abroad, were delayed for substantial periods of time. Submission of individual- and institutional-centric research projects reduced or closed. In some cases, fellowships to project staff were not disbursed. Many national and international academic events were cancelled. The number of webinars significantly increased but failed to achieve one-to-one academic discussion and academic exchanges in a true sense. There were not visits of scientists to overseas HEIs and laboratories under academic exchange programmes. Expensive instruments and equipment were damaged in the absence of trained people to maintain and run these machines on a regular basis. Many chemicals with limited shelf-lives had to be thrown away. Many scientists had deep concern that even after the situation returns to normal, labs will not be able to function normally, and it may take at least three to six months bringing it to normalcy (The Wire-Science, April 2020). These consequences have had a negative impact on the psychology of researchers.

## 6.9 POST-COVID RESPONSE FOR HEIs

The impact of the first and second wave of COVID-19 has been over but the third wave of COVID-19 was not at all severe as compared to the earlier two waves, although the number of COVID-related deaths has been substantially reduced by increasing vaccinations and improving health infrastructure (SBI Report – June 1, 2021). Educational and research institutions, industrial and government bodies need to join hands together in a mutually helpful and congenial environment to help each other (Figure 6.3). Some initiative in this regard was taken up by O.P. Jindal Global University and The Association of Indian Universities who have jointly developed a COVID-19 Response Tool Kit (HEI-CR Tool Kit) document for Indian HEIs, released during August 2020. The same has provided

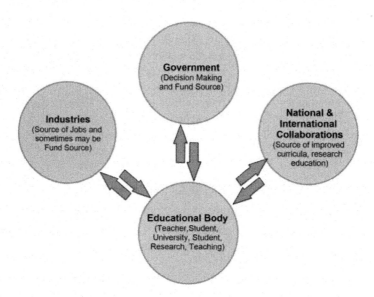

**FIGURE 6.3**   Illustration of interactive response mechanism.

guidelines for filling gaps in studies, delays in sessions, etc., along with mention of the importance of guest lectures, etc., to keep students and teachers motivated.

## 6.10   CONCLUSIONS

While we know it is difficult to predict the future impact of COVID-19 on HEIs, we do know that online teaching-learning, examination, and evaluation is a viable response to dealing with the pandemic and increasing global collaboration to help the educational community keep going.

## ACKNOWLEDGEMENTS

We extend our sincere thanks to Padma Shree Professor R. C. Sobti who gave us the opportunity to contribute this chapter. The chapter is a kind of compilation, review, and critical evaluation of COVID-19 information mostly available on various websites and print media reports, with regard to higher education institutions of India. Omission of any relevant reference may have been done inadvertently. We especially thank Pranjal Yadav who digitized the illustrations in this chapter.

## REFERENCES

Adam, D. (2020). Special report: The simulations driving the world's response to COVID-19. *Nature*, 580(7802), 316.

Arora, V. P. S. (2016).' Higher education in India: Status, issues, and chalenges. In: Arora, V. P. S., Pathak, D., Goswami, V. K., Agarwal, S. S., Marwah, S. (Eds.), *Higher Education: Leadership and Management. Centre for Education Growth and Research (CEGR)*. T Bharti Publication, New Delhi, pp. 1–17.

COVID-19 Response Toolkit for Indian Higher Education Institution, Institutional Resilience for Academic Planning & Continuity (August 2020).

COVID-19: Lockdown Across India, in Line with WHO Guidance. *UN News*. 24 March 2020.

Edmund, A. (2020). www.universityaffairs.ca/features/feature-article/the-tale-of-two-pandemics/

Gettleman, J., Schultz, K. (24 March 2020). Modi Orders 3-Week Total Lockdown for All 1.3 Billion Indians". *The New York Times*. ISSN0362-4331.

https://en.wikipedia.org/wiki/COVID-19_pandemic_lockdown_in_ India

https://mhrd.gov.in/sites/upload_files/mhrd/files/lu2929.pdf

https://science.thewire.in/the-sciences/coronavirus-lockdown-india-research-laboratories-cell-lines-animals-physics-astronomy/

https://timesofindia.indiatimes.com/readersblog/unplugged/the-walk-to-home-or-god-19828/

Kumar, S. (2016). Documentation and requirements for accreditation and ranking of Higher Education Institutions. In: Arora, V. P. S., Pathak, D., Goswami, V. K., Agarwal, S. S., Marwah, S. (Eds.), *Higher Education: Leadership and Management. Centre for Education Growth and Research (CEGR)*. Bharti Publication, New Delhi, pp. 262–274.

Morgan, K. R. (2002). *Blended Learning: A Strategic Action Plan for a New Campus*. Seminole, FL: University of Central Florida.

Reay, J. (2001). Blended learning – a fusion for the future. *Knowledge Management Review*, 4(3), 6.

Rooney, J. E. (2003). Blending learning opportunities to enhance educational programming and meetings. *Association Management*, 55(5), 26–32.

SBI Report Link (Published June 1, 2021). https://sbi.co.in/documents/ 13958/10990811/010621-Ecowrap_ 20210601.pdf/fe475ae5-b0ce-c453-f090-62f0cf548f70?t=1622532682984

Takwale, R. (2003). Challenges and Opportunities of Globalization for Higher Education in India – Alternatives through e-Education, UGC Golden Jubilee Lecture Series. www.ugc.ac.in/ oldpdf/pub/lectures/ugc_ pro2.pdf

The Indian Express: Unlock 5.0 Guidelines. https://indianexpress.com/ article/india/unlock-5-coronavirus-guidelines-6619158/

www.google.com/search?q=AI+system&rlz=1C1LENP_enIN481IN481&oq=AI+system&aqs=chrome..69i57.2593j0j8&sourceid=chrome&ie=UTF-8

www.livemint.com/education/news/india-s-higher-education-student-population-grows-by-8-lakh-hrd-ministry-1569081600712.html

www.researchgate.net/post/Can_technology_replace_a_teacher

www.swayamprabha.gov.in/index.php/about

www.ugc.ac.in/oldpdf/pub/he/HEIstategies.pdf

www.ugc.ac.in/page/C-E-C.aspx

www.ugc.ac.in/pdfnews/9208605_Brochure-(National-Convention-on-Digital-Initiatives-for-Higher-Education).pdf

www.ugc.ac.in/ugcpdf/740315_12FYP.pdf

# 7 Academic Administration

## *Post-Pandemic Challenges in Science Education*

*Ashish Kumar Lamiyan[1], Ramkesh Dalal[1],
Neelima R. Kumar[1]\*, and R. C. Sobti[2]*

[1]Department of Zoology, Panjab University, Chandigarh, India
[2]Department of Biotechnology, Panjab University, Chandigarh
\*neelimark6@gmail.com

## CONTENTS

## 7.1 INTRODUCTION

The term pandemic is derived from the Greek word *pan* meaning "all" and *demos* meaning "people." It is an epidemic of an infectious disease that has spread over a large region, involving several countries, continents, or worldwide, affecting a substantial number of people.

The most fatal pandemic recorded to date was the Black Death (also known as the Plague), which killed an estimated 75–200 million people in the 14th century. Another disastrous pandemic was the Spanish flu of 1918, while important current pandemics include HIV/AIDS and COVID-19.

### 7.1.1 CAUSE OF PANDEMIC

It is a well-known fact that global warming is leading to unprecedented climate change, which is resulting in disturbed weather patterns such as prolonged rains in some regions, draught in others, and totally changed climates in still others. These climate changes are conducive to the sustenance of a diversity of vectors for longer periods of time, allowing them to multiply rapidly, and also creating situations suitable for the survival of new vectors.

This is responsible for an increase in the prevalence of infectious diseases – both new and recurring. In 2016, the United Nations Environment Programme published a report called: "UNEP FRONTIERS 2016 REPORT." In this report, a special chapter was dedicated to zoonotic diseases (i.e., diseases that pass from animals to humans). It was written therein that deforestation, climate change, and livestock agriculture are among the main causes that increase the risk of such

DOI: 10.1201/9781003358916-7

diseases. The more civilized humans become, building cities and forging trade routes to connect with other cities, and waging wars with them, the more likely pandemics become (United Nations Environmental Programme, 2016).

### 7.1.2 PANDEMIC MANAGEMENT

As the pandemic continues to bewilder the world, special protocols and procedures are needed. The basic strategies in the control of a pandemic are containment and mitigation. Containment may be undertaken in the early stages of the outbreak, including contact tracing and isolating infected individuals to stop the disease from spreading to the rest of the population, while mitigation includes other public health interventions in infection control, and therapeutic countermeasures such as vaccinations (Lipton & Prado, 2020).

A broad group of the so-called non-pharmaceutical interventions may be taken to manage the outbreak. In a flu pandemic, these actions may include personal preventive measures such as hand hygiene, wearing face-masks, and self-quarantine; community measures aimed at social distancing such as closing educational institutions, shopping malls, gyms, salons, and cancelling mass gatherings, community engagements, and religious gatherings; and environmental measures such as cleaning of surfaces, sanitization, and disinfection (Ahmed et al., 2020).

### 7.1.3 THE CURRENT PANDEMIC

A new strain of coronavirus, which was first identified in Wuhan, Hubei province, China, in late December 2019, caused a cluster of cases of an acute respiratory disease, which is referred to as coronavirus disease 2019 (COVID-19) (Wu et al., 2020). According to media reports, more than 200 countries and territories have been affected by COVID-19, with major outbreaks occurring in central China, Iran, Western Europe, the United States, Brazil, and Russia. On March 11, 2020, the World Health Organization characterized the spread of COVID-19 as a pandemic (Jiang et al., 2020).

> When COVID was intense in mid-March of 2020, governments the world over were reacting to it. The virus was new and its behaviour was so unpredictable. The pandemic affected universities and institutes across the globe and closed down education and research systems (United Nations Educational, Scientific and Cultural Organization, 2020). Many institutions put in place online teaching options, though some reported a lack of proper resources and not all the professors being able to transfer teaching online. One of the biggest initiatives launched by the Baylor Institute in the United States is professional development on virtual teaching. Prior to March, only 8% of Baylor faculty had taught online before. In the program faculty with previous experience teaching online mentor other professors to create more robust online learning experiences for students (Hillier, 2018).

As of June 3, 2020, the Chronicle of Higher Education had tracked 860 institutions' plans for opening in the fall, with 67% planning to resume, 18% waiting to decide or considering a range of scenarios, 8% proposing a hybrid model, and 7% planning for online. One thing that is certain is that higher education will look radically different in the coming times with smaller class sizes to more technology.

Some institutes are anticipating offering the majority of their courses face-to-face. Many colleges and universities are providing faculty training on effective online learning, retrofitting classrooms to allow for social distancing, modifying course schedules, and rethinking on-campus housing and dining. Many institutions that are preparing for in-person instruction are making virtual alternatives available for students with health risks. Others that are planning to offer the majority of classes online are figuring out how to provide limited face-to-face instruction for disciplines that require hands-on training (Hillier, 2018). But everything from the academic experience to athletics is going to look different. "Until you have the therapeutics, until you have a vaccine, until those things are in place, it's going to be different for all of us around the world," he says. "And universities are going to be no different. We need to learn to live with it. You need to learn to coexist with it. That's exactly what we're going to do."

In preparation for an extreme situation the University of Tennessee, Knoxville (UTK) set aside 170 beds that could be used as isolation units in case students need to be quarantined. They also ask students in the dorms to take their temperature in the morning and plug it into an app to determine whether or not they should go to class.

## 7.2 INTRODUCTION OF ONLINE EDUCATION

Online education in its various modes has been growing steadily worldwide due to the confluence of new technologies, global adoption of the Internet, and intensifying demand for a workforce trained periodically for the ever-evolving digital economy. Online education is on track to become mainstream by 2025 (Hunt, 2011). The country-level factors that impact quantity and quality of online education have been influencing almost all aspects of our lives: the way we work, interact with others, process data into information, analyze and share information, entertain ourselves, and enjoy tourism. E-evolution or e-revolution has witnessed e-mails, e-commerce, e-government, and now e-education. E-education or online education is changing the way we approach teaching and learning. Changes in education delivery models have been rapid and transformational. As institutions worldwide adapt to these changes, a very dynamic education landscape has generated immense interest among researchers, educators, administrators, policymakers, publishers, and businesses (Palvia, 2013). Today's "online" or "blended" learning started in the 1990s with the advent of the Internet and the World Wide Web and reached individuals in remote locations, or those who wanted the convenience of eliminating travel time. As information and communication technologies have kept advancing, online education has become more feasible technologically, economically, and operationally. Incentives for universities to offer online programs include financial constraints and rewards (e.g., reduced infrastructure for classrooms, offices, cafeterias, dorms, and libraries), increase in non-traditional students that are working full-time, and advanced state of technology making it easy to implement. This rich and diverse history of online education has produced a substantial body of research, examining different aspects of online education. Many conferences and journals have had themes and special issues focusing on online education (Hillier, 2018; Leelavathi, 2020).

## 7.3 TRAINING FOR TEACHERS AND STUDENTS

Low- and middle-income countries have to face extra challenges in the sudden transition to change in already established working education system. The main concern in this regard is for students who are unable to access digital classrooms as they live in areas without fast, reliable, and affordable broadband, or where students have no access to laptops, tablets, smartphones, and other essential hardware. Approximately one-quarter of students who have access struggle to continue their working routines and stay updated during lockdown. On the other hand, students from poorer households in remote regions were not able to travel to the nearest places where they could have access to the internet to download course materials. There is a very long way for India to go in order to get the technology very effectively to areas that are underserved. Also, it is essential for government and funding bodies to ensure that in an emergency situation all students and universities are eligible for the same kind of facilities (Leelavathi, 2020).

In poorer countries, the public funding that universities rely on is under threat because economies have crashed during lockdowns. The governments in these countries and other countries need to realize that the impact of such decisions will fall disproportionately on the poorer and less facilitated students.

The growth of online learning during the last decade has been remarkable. While in 1998 there were too few students enrolled online to count, according to a survey of more than 2500 institutions, by 2009 more than 5.6 million students were taking an online course. Nearly 30% of students were taking a course online. The same study also found the percentage of enrollment growth was 21%, while overall growth in higher education was only 2%. Moreover, the 21% growth rate for online enrollments far exceeds the less than 2% growth of the overall higher education student population. These numbers indicate that online learning has become an important mode of delivering instruction in higher education (Ray & Subramanian, 2020).

Although the numbers of students taking online courses are growing, research indicates that the students are in many ways the same students who take courses offline (Doyle, 2009). Students tend to be relatively similar when comparing race, gender, socioeconomic status, and physical distance from the institution. Students who take online courses tend to be slightly older than those students taking all courses offline (Doyle, 2009). Several important studies have documented that these students have good learning outcomes in online courses. Such research most frequently compares online to offline courses in experimental or quasi-experimental studies (Bernard et al., 2009; Gunawardena & McIssac, 2004; Lockee et al., 2001). Technology enables social integration to shift from face-to-face communication to more disembodied forms of communication, so participants in the online environment can engage outside of one another's presence (co-presence). Several students were grateful for the opportunities that online learning presented them for access to higher education. Interestingly, the most common potential barrier to educational attainment that students mentioned was family, and that also was the factor that made them most grateful for the opportunity and the experience. The challenge of managing time seemed to arise from the amount of communication during the attending of the online course. Their sense of their own abilities in this area influenced their experiences. Students in several cases felt some responsibility for course outcomes seemed to have more positive experiences (Palvia et al., 2018).

Teachers also had a strong influence over student experience, in large part through their accessibility and through their efforts to provide opportunities to connect with peers. The accessibility of teachers is an important aspect that strongly influences the student experience. Unfortunately, not all students had a positive experience with their teachers, and so their online experiences suffered. Professors who were more comfortably available and easily accessible led to better experiences of students.

Sometimes students were unable to understand the expectations of the professor's assignment due to inaccessability. Some students, on the other hand, felt that they did not have a chance to make connections with other students in the online courses, which left them feeling isolated.

## 7.4   CREATION OF MODULAR APPS

The Indian government has launched the National Mission on Education through Information and Communication Technology (NMEICT) to effectively increase the potential of information technology in the teaching and learning process. The end goals of this mission are to provide a quality level of education in the country. The main focusing factors include both pros and cons that likely determine success of online education in the Indian context.

India is a developing country and represents a better position for the adoption of online mode of education regarding the strata of its development among the other developing countries as mentioned by KPMG India and Google (Bansal, 2017). As per the study made by KPMG India and Google, the online education system in India is expected to grow to US $1.96 billion with around 9.6 million users by 2021, which currently with an average of 1.6 million users is at US $247 million. Both the agencies believe that the major driver leading to an increase in online/blended education in India is the phenomenal growth recorded in Internet and an escalating smartphone usage by individuals (Bansal, 2017). The cause for the attraction towards this mode is determined by its low cost, digital-friendly government policies, and escalating demand by working class and job-seekers. Some of the government initiatives include Digital India and Skill India launched to spread digital literacy in Indian grounds. More similar leading programmes are e-Basta (schools' books in digital form), e-Education (all schools connected with broadband and free WiFi) in all schools, development of pilot

MOOCs (Massive Online Open Courses), NandGhars (digital tools as teaching aids), SWAYAM (MOOCs based on curriculum taught in classrooms from 9th class until post-graduation), and India Skills Online (learning portal for skill training). Clearly, these government programmes go a long way in reaping benefits of online education in the face of the challenges in India.

## 7.5  DIFFICULTIES FOR SCIENCE STUDENTS

The current situation has imposed a huge impact on science education. All agencies and institutes sponsoring research are demanding accountability of the productivity of researchers. Many research projects have been interrupted because of workspace and equipment unpreparedness. On the other hand, this has created opportunities for assessing data-related anthropogenic effects, containment measures, and other types of environmental adaptation-related activities.

Virtual labs were introduced keeping in mind prior validation and testing the hands-on lab. This approach under normal circumstances makes it possible to reduce the damage that can be caused in the real system (hand-on labs) due to common and frequent errors. Students can repeatedly see and remember the main elements that are necessary and can understand how the different components work (Bailenson, 2018).

This versatility allows students to know, analyze, and evaluate different communication alternatives to continue the research. When students are pursuing research in subjects like engineering and life sciences the best way to learn is by practicing only. The advantages of using virtual mode include enhancement of concept learning beyond the classroom; virtual labs can use an assortment of machines or processes. This will add diversity to each laboratory session and which will enable student learning; and skill strengths during different practices, allowing acquisition of competencies that are usually faced at the beginning of research career. During the last few months, both remote lab and virtual lab modes have been used in courses for students using practical systems of research.

## 7.6  HYBRID MODELS

Some colleges are considering a middle ground between face-to-face and virtual learning. Boynton from the Beloit Institute says his institution ultimately wants to bring students back to Beloit for face-to-face instruction, but to be able to respond to changing circumstances they adopted a new modular approach that will allow for maximum flexibility.

"The American public is not served well by being either online or face-to-face," he says. "There's no way either one of those possibilities is going to happen. It's going to be some combination of the two" (West, 2020). Beloit landed on its current plan as it looked at ways of increasing flexibility and decreasing risk. The model breaks the semester into two halves and students take two courses during each "mod." "Generally taking four online courses at the same time is debilitating, both for faculty and students," he says. "Splitting it up into two different 'mods' …takes some of the heat off of having to juggle four classes simultaneously online" (Bailenson, 2018). The modified schedule also makes it easier to pivot back to online learning should a subsequent COVID-19 outbreak occur. It also reduces the number of times students switch classrooms, which makes cleaning easier. Given all of the uncertainty around COVID-19, the plan allows flexibility in the things it does have control over: the academic calendar and delivery of curriculum.

## 7.7  CONCLUSIONS

Laboratory experiences are an essential part of science education enabling students to reinforce concepts and develop and practice skills based on the theories. Due to Covid, different types of

ONLINE      OFFLINE

MODE       MODE

OR

The difficult decision!

**FIGURE 7.1**   Confusion in making decisions.

laboratories in universities around the world have initiated the use of different types of platforms for continuation of research and science education. Hands-on labs rely on technology to facilitate student learning, which makes the learning processes more efficient. It has been observed that the pandemic has inculcated a sense of collaboration and sharing of technology among organizations and students to learn concepts and practice skills and apply them during the development of projects. All organizations have one aim: how can technology be utilized to improve students' skills and help them to acquire knowledge better, so they can enter the workforce trained? There is a great need for new teaching paradigms that encourage active learning and the development of skills based on challenges. The availability of any type of laboratory is not a sufficient condition to ensure success in the teaching/learning process. Any type of laboratory offered as a stand-alone setting, without connection to adequate learning material, usually leads students to the use of a trial and error strategy, which has a lower learning impact. Students have greater acceptance and more interest in working with the new technologies and exercising hands-on approaches. A blended/hybrid approach (i.e., combination of hands-on and virtual labs) may be a better proposal for experimental/ research education instead of a single laboratory approach. A blended/hybrid laboratory can offer the opportunity to exploit the key features of each type of laboratory based on the desired objectives. Hands-on labs could be used in the first stage to build confidence in the virtual technology that can be used further for better research and development. Prof. D.P. Singh, Chairman, University Grants Commission, in an interview with Rekha Dixit, Senior special correspondent, The Week, said that the UGC recognised the importance of traditional face-to-face teaching, but the recent developments in the tools of information and technology encourage adoption of these tools in order to prepare ourselves for challenges such as the present one. UGC would therefore like to promote a hybrid approach. In the end one can only say that every place is different. It depends on where your universities are and the rules and laws of that particular state. Each university system needs to think about what its particular situations are, and then come up with contingency plans…and be prepared for a blind turn (Figure 7.1)!

## REFERENCES

Ahmed S, Duan C, Huremović D, Hussain S, Khan S, Levin J, Victor GS (2020). Psychiatry of pandemic. *Psychiatry* 1: 3.

Bailenson J (2018). *Experience on demand: What virtual reality is, how it works, and what it can do.* WW Norton & Company.

Bansal S (2017). How India's ed-tech sector can grow and the challenges it must overcome. VC Circle. www.vccircle.com/the-present-and-future-of-indias-online-education-industry

Bernard R, Abrami P, Borokhovski E, Wade CA, Tamim R, Surkes M, Bethel E (2009). A meta-analysis of three types of interaction treatments in distance education. *Review of Educational Research*, 79(3): 1243–1289.

Doyle W (2009). Online education: The revolution that wasn't. *Change* 41(3): 56–58.

Gunawardena, C, McIssac M (2004). Distance education. In D. H. Jonassen (Ed.), *Handbook of Research for Educational Communications and Technology* (2nd ed., pp. 355–396). Mahwah, NJ: Erlbaum.

Hillier M (2018). Bridging the digital divide with off-line e-learning. *Distance Education* 39(1): 110–112. doi:10.1080/01587919. 2017.1418627

Hunt C (2011). *National strategy for higher education to 2030*. Department of Education and Skills.

Jiang F, Deng L, Zhang L, Cai Y, Cheung CW, Xia Z (2020). Review of the clinical characteristics of coronavirus disease 2019 (COVID-19). *Journal of General Internal Medicine* 35: 1–5.

Leelavathi R (2020). Digital education in current scenario. *CLIO: An Annual Interdisciplinary Journal of History* 6(9): 645–656.

Lipton A, Lopez de Prado M (2020). Mitigation Strategies for COVID-19: Lessons from the K-SEIR Model. Available at SSRN 3623544.

Lockee B, Moore M, Burton J (2001). Old concerns with new distance education research. *Educause Quarterly* 24(2): 60–63.

Palvia S, Aeron P, Gupta P, Mahapatra D, Parida R, Rosner R, Sindhi S (2018). *Online education: Worldwide status, challenges, trends, and implications.*

Palvia SC (2013). E-evolution or E-revolution: E-mail, E-commerce, E-government, E-education. *Journal of IT Case and Application Research*, Editorial Preface Article 15(4): 4–12.

Ray D, Subramanian S (2020). India's Lockdown: An Interim Report (No. w27282). National Bureau of Economic Research.

United Nations Environmental Programme (2016). UNEP FRONTIERS 2016 REPORT: Emerging Issues of Environmental Concern (accessed March 19, 2020).

United Nations Educational, Scientific and Cultural Organization. COVID-19 Educational Disruption and Response. (2020). https://en. unesco.org/themes/education-emergencies/coronavirus-schoolclosures (accessed March 19, 2020).

West C (2020). Academic Planning in a Pandemic. *AGB Magazine*. 28(4): 34–38.

Wu F, Zhao S, Yu B, Chen YM, Wang W, Song ZG, Yuan ML (2020). A new coronavirus associated with human respiratory disease in China. *Nature* 579(7798): 265–269.

# 8 Family Life and Education during COVID-19 Pandemic in India

## Some Observations

Jagdish C. Mehta

Department of Sociology, D.A.V. College, Chandigarh, India
jagdishdavc@gmail.com

## CONTENTS

## 8.1   INTRODUCTION

Pandemics in previous eras like the Spanish flu, influenza, smallpox, cholera, swine flu, SARS, and H7N9 had a tremendous impact on society. Presently, the world is dealing with a severe and challenging health crisis with the newest pandemic, COVID-19, which started in December 2019 from Wuhan, China and spread to the whole world. The outbreak was declared a Public Health Emergency of International Concern in January 2020 and a pandemic in March 11, 2020 *(Saji et al., 2021)*. It has occurred in an age when the global economy is more integrated than ever, and people, goods, money, and ideas move faster than they have ever done before. Consequently, the effect of a microbial disease with intercontinental spread and fatal potential has affected societies all over the world more than at any time in the past. The coronavirus pandemic has become much more than a global health crisis. It has become a human, economic, and a social crisis. In addition to the disease that has cost us so many lives, it has caused havoc in various sectors of life – economic, social, cultural, and behavioral.

Governments all over the world including India have taken strict precautionary measures to reduce the spread of coronavirus such as social distancing, hand hygiene, wearing face masks, and closing schools, colleges, transport spaces, restaurants, shopping malls, and other places where people might gather. The Government of India ordered a nationwide lockdown for 21 days starting from March 16, 2020. Most citizens were prohibited from leaving their homes except for healthcare workers, police, and workers involved in other emergency services. Hence, the movement of the entire 1.3 billion population of India was limited as a preventive measure against the COVID-19 pandemic. The lockdown was further extended until May 3 and thereafter to May 31 by the National Disaster Management Authority. The Ministry of Home Affairs announced that the ongoing lockdown would be further extended until June 30 in containment zones, with services resuming in a staggered approach.

The enforced restrictions of staying home brought about a radical change in people's livelihoods and lifestyles. As a consequence, lifestyles, behavioral patterns, family life, and work culture have undergone a significant change. The educational system too has faced many challenges and has undergone a radical shift. Thus, the present study is going to examine the positive and negative impacts of COVID-19 and its precautionary measures on family life and education.

## 8.2   OBJECTIVES AND METHODOLOGY

The present study aims to analyze the impact of COVID-19 pandemic on two pertinent social aspects, family life and the education system, as both are intertwined and have a bearing on each other. With regard to family, this study engages with different aspects of a family being impacted by COVID-19; the experiences of family members, the quality of relationships, division of labor, leisure activities, professional work, family conflict, etc. In the educational front, the study engages with the emerging methods of teaching and learning with the use of digital technology. This study also examines the challenges faced by teachers and students due to the use of technological alternatives in various aspects of education like teaching-learning, evaluation, assessment, and examination.

The study is primarily based on secondary sources, review of existing studies and personal observation, interactive sessions with the people across groups based on gender, caste, economic status, etc., focused group discussion among acquaintances, and 40 students of M.A. Sociology at D.A.V. College, Chandigarh.

The analysis is divided into the following two sections. Section 1 deals with the social impact of COVID-19 breakdown on family life and the associated indicators and Section 2 looks into the impact on education and emerging patterns of teaching, learning, assessment, and evaluation methods along with the challenges faced by teachers and students.

## 8.3   FAMILY LIFE UNDER THE SHADOW OF COVID-19 PANDEMIC

This section looks at the opportunities and challenges families in India have faced during the COVID-19 pandemic. COVID-19 is not only a severe health crisis that has taken many lives, but it is also daunting socio-psychological and economic challenge that many families are coping with. Almost all families have had to face many challenges due to disruption of trade, connectivity, and social services. This disruption, on the other hand, might have brought with it many opportunities to resolve the deterioration of family relationships and interactions *(Addati et al., 2018)*. Wright and *Leahey (2013)* are of the view that serious illness and life challenges have greater bearing on the family unit, and reciprocally, the functioning of the family unit influences the health and well-being of each family member. This holds true for the current COVID-19 pandemic, which has uniquely affected families by disrupting routines, changing relationships and roles, and altering usual child-care, school, and recreational activities. One cannot deny the negative impact of COVID-19 in the structure of the family, but there have been positive changes also in the day-to-day activities of the family as a result of lockdowns.

### 8.3.1 Negative Impact of COVID-19 on Family Life

#### 8.3.1.1 Family Relationships are under Pressure Due to the Additional Responsibility of Taking Care of Children, Aged, and Other Vulnerable Relatives

Lockdown policies forced citizens to cooperate with one each other at various levels to control the pandemic spread. Here, the burden to prevent the disease fell mainly on families. On the one hand, family members had to cooperate with the government in respecting lockdown measures; and on the other they were burdened with additional responsibilities in their households as the usual divisions between work, home, and school became blurred.

Apart from taking more preventive measures, most family members were struggling with learning the latest methods to survive in a world of pandemic like e-conferencing, online shopping, and e-learning as well as performing their household chores and livelihood activities at home or in confined spaces. Preventive measures to control the spread of the coronavirus had disruptive effects on relationships in general and family relationships specifically. Families reported a loss of a sense of a community feeling in response to quarantine/lockdown measures. Other tangible losses were income, access to resources, access to health care as well as sources of recreation like planned activities or celebrations, the opportunity to host or attend others' events of social ceremonies, weddings, birthday parties, retirement functions, etc., while also increasing the time or frequency at which family members communicate effectively to resolve many issues. This resulted in an entirely new form of effect on family unity that has never been discussed before (Woodruff, 2020).

The lockdown measures instituted in India have also invited vulnerability and risk within families. Work from home and closure of schools lead to distress in many families not accustomed to being so closely confined for a long time period. Moreover, as a result of the COVID-19 crisis, outside support given to families who provide long-term care for an ill parent, partner, or child is lost. Families with a child who requires specialized care and guidance have had to care for their child 24 hours a day without the outside guidance provided by medical institute, daycare, or special services. Families who care for a father, mother, or partner with dementia or other serious illness have had to manage without day care or institutionalized assistance.

In the absence of school, childcare, extra-curricular activities, and children's social support networks have been broadly disrupted. Stress from COVID-19 has been compounded by additional responsibilities for parents as they adapt to their new roles as educators and playmates while balancing full-time caregiving with their own stressful changes to work, financial, and social situations.

School closures have created a family environment where children are rarely allowed to leave the house and may be confronted by the vulnerabilities of a family member's addiction, aggression, and violence. Children of divorced co-parents are suddenly being refused alternating parental care because one of the parents now works from home and cannot provide childcare. The increasing stress and vulnerability have also been observed amongst divorcees and as well as people who are not married as they feel insecure and lonely.

### 8.3.2 Reshaping Family Dynamics

Lockdown and social distancing have changed the whole family dynamics. People have been cut off from friends and broader societies as a result of these initiatives (Woodruff, 2020). Stressful experiences like physical distancing and orders to stay at home have intensified threats of growing domestic violence such as physical, emotional, and sexual abuse. Moreover, in cases where family violence already existed before the COVID-19 pandemic, the lockdown and isolation may have worsened the situation. Many studies across the world show that the accumulation of pandemic stress, in addition to the stress of working from home, as well as the lengthening of school closures and social isolation, has resulted in an increase in violent behaviour among some family members. This has led to a rise in strain amongst families and marital conflicts in certain families and communities *(Buheji et al., 2020; Liu, 2020; Power, 2020; Woodruff, 2020)*. The COVID-19 pandemic in China prompted unprecedented divorce demands, prompting the government to create a 30-day cooling-off period mandatory in various provinces. During the lockdown, the tensions in China resulted in an increase in domestic abuse incidents. In the first week of the pandemic, more than 94,000 cases of family violence were recorded in South Africa, while France saw a 36% increase in some form of family violence during the same time span. In the first few weeks of the lockdown, Australia saw a 75% surge in domestic violence. In India, figures on domestic violence doubled within the first week of the lockdown. Many families and partners found themselves navigating new challenges as even the less serious cases of dispute also aggravated the tensions in families *(UN, 2020)*.

Multiple reports in India speak about the surge in violence during the COVID-19 pandemic. The National Commission for Women has recorded a two-fold increase in gender-based violence across the country, with the group receiving 257 calls in the final week of March as opposed to 116 calls in the first week. The realities of these numbers loom several times greater than their value with 99% of sexual assault cases in India going unreported, and a woman being 17 times more likely to be assaulted by her husband. The situation has become so grave that the National Legal Services Authority decided to open an online legal services call line by eminent women lawyers for each district.

Even men have become victims of abuse. The loss of job or salary cut because of the lockdown has put men in a situation where they have to bear the wrath of family members or their spouses. Due to the patriarchal mindset prevalent in our societies, violence against husbands is often neither believed nor recognized. Thus, men in general and husbands in particular suppress the feeling and emotions of abuse by their wives at home or bosses at the workplace.

This increase in family violence may be primarily due to large-scale lockdowns that aggravate family conflicts, economic distress, and tension caused by the pandemic among family members, and inadequate support for victims of family violence during the pandemic. Family violence has become an important social issue that needs to be dealt with properly and swiftly during the ongoing COVID-19 pandemic. Both governmental and non-governmental organizations (NGOs), as well as agencies and service sectors, need to be aware of the needs of family violence victims and provide appropriate and immediate assistance to the victims during the pandemic.

### 8.3.3 IMPACT ON VULNERABLE FAMILIES

The families of migrant workers and labourers have faced multiple hardships due to the COVID-19 lockdown. Migrant workers in both urban and rural areas have had to face economic hardships and distress. Due to the sudden lockdown, migrant workers were left with no option; they had to move back to their hometowns. They were offered no support by the government. Instead, they were met with apathy, violence, and a burdensome attitude. Due to lack of income, these families were not even able to get one square meal a day. The returning migrant workers reported that they were stranded in their destination cities for one to two weeks before they get back to their home districts. During the time they were stranded, many of them had to vacate their housing, including rental housing and worksites. They needed to vacate the rented houses due to forced eviction by the landlord, loss of job, closure of factory, and inability to pay the house rent due to loss of livelihood. The workers rarely received any form of assistance (either for travel/food/housing) from the government (Kasarla Harshitha, 2020, The Wire). Thus, the whole family faced the trauma of homelessness, joblessness, insecurity, fear, and the wrath of their own family members at their household in their hometowns.

In summary, after months being together or displaced due to COVID-19 lockdown, some families started to experience symptoms of burnout. The above analysis shows that inescapable togetherness contributed to fractures in relationships. Some family members also experienced physical or emotional exhaustion and thus felt that they were not able to manage family affairs effectively. Therefore, in order to mitigate burnout, steps should be taken to such as by sharing responsibilities, positive thinking, and exploring advantages of being together to help families find balance during this stressful time.

## 8.4 POSITIVE IMPACT OF COVID-19 ON FAMILY LIFE

Although lockdowns have kept people indoors certain positive social effects have been seen as well. Familial bonds strengthened, and people became more creative. It gave people time to rethink the effects of our actions on the environment. It stirred people to introspect, to go within themselves and find their purpose in life.

### 8.4.1 OPPORTUNITY TO SPEND QUALITY TIME WITH FAMILY MEMBERS AND REJUVENATE RELATIONSHIPS

The lockdown period gave people an opportunity to take a pause, refresh, and reconnect not only with loved ones but with one's self too. Families have been forced to spend endless time together and have begun to realize the value of these relationships. Conjugal relations have been rekindled and parental bonds have been rejuvenated, albeit with exceptions of course!

Keeping in touch with one's extended family and friends is crucial to overall health and happiness. Keeping a good attitude during physical distancing and lockdown can be a perfect time to restore, improve, or sustain family ties that have deteriorated or become dormant over the years *(Azarian, 2020)*. Family members can choose to do things they used to do together, bringing back memories of what they used to do, or they can do something different that they have never done before. Spending quality time with family members encourages them to see time as an opportunity for all family members to work together through difficult times. This contributes to the development of stronger ties and resilience. Quality family time necessitates devoting some of our time to one-on-one conversations with family members in order to reinforce bonds *(Roshgadol, 2020; Szabo et al., 2020)*.

*Buheji et al. (2020) and Szabo et al. (2020)* report that indoor sports games are critical in reinforcing positive family attitudes and behaviors. Similarly, playing outdoor games or doing physical activity together regularly while following Covid norms, through sharing of experience, or preparing outdoor snack meals together, etc., can turn everyday moments into quality time.

It has been observed during focused group discussion with acquaintances that family life has changed positively in many ways. Lockdown has given people a golden opportunity to spend time with their families. Family members, especially adults, are used to remaining busy and engaged in various pursuits such as careers and businesses, and lockdowns provided an opportunity to slow down.

Parents have a renewed focus towards family and personal relationships, with many reporting the desire to reach a better balance between work and family life post-lockdown, with some parents actively looking for new employment or more flexible working patterns.

It has been observed that spending more time together as a family is beneficial for many children. School closure may have resulted in parents taking greater care in preparing their children's meals and helping them with schoolwork. At the same time, many parents prioritized their child's wellbeing above educational attainment during this time, leading to reports of positive child wellbeing and outcomes.

### 8.4.2 SELF-REREALIZATION AND LEARNING NEW SKILLS

Lockdown has made people reflect and consider deeper insights. It has shaken people and made them question the purpose of life, their existence. Members of the focus group shared that how lockdown provided the opportunity to experience peace and quiet due to there being no traffic, again hearing birds chirping and other sounds of nature, to gaze at trees from home and wonder about the sheer beauty and complexity of nature. Other people worked on improving their skills in meditation and finding their talents in arts like singing, drawing, gardening, cooking, etc.

### 8.4.3 Value of Domestic Labor Has Been Redefined

Household chores are now more commonly equally shared amongst family members, resulting in a camaraderie developing. It has been observed that now husbands are sharing in household chores with wives and in return are also taking extra responsibility of household chores in the absence of maids by taking care of children and elders.

## 8.5 EDUCATION UNDER THE SHADOW OF COVID-19 PANDEMIC

Education plays an important role in bringing about social progress, economic growth, and egalitarianism and thus paving way to enrich lives (Mehta, 2014, 2017). The COVID-19 pandemic has wreaked havoc on the world's educational systems, affecting nearly 1.6 billion people in more than 200 countries. About 94% of the world's student population was affected (UNESCO, 2020b). All educational centers and other learning spaces were shuttered, with classes suspended and examinations at schools, colleges, and universities including entrance tests postponed. Thus, lockdowns destroyed the schedules of students, teachers, and families.

The need of the hour was to innovate and implement alternative educational systems and assessment strategies. The COVID-19 pandemic compelled the world to use digital technology in education to ensure learning continued (Dhawan, 2020). The Government of India took measures such as developing online learning platforms to ensure that the academic activities of schools and colleges continued during lockdowns.

Initially, educational stakeholders were critical of education through digital media, but later realized the value of innovative alternatives in education during the extended pandemic. Thus, COVID-19 posed many challenges and also provided some opportunities for educational institutes (Pravat, 2020a). This section of the chapter evaluates the impact of the COVID-19 pandemic on teaching and learning in India.

Closure of educational centres not only influences learners, teachers, and families, but also affects the social and economic status of many. Emerging educational requirements in response to COVID-19 have highlighted numerous global issues such as digital learning, food insecurity, childcare, internet, healthcare, disability services, etc. The subsequent impacts are higher on poor children and their families, resulting in discontinuity in learning, adverse physical and psychological health issues, and childcare and associated economic costs for families who are unable to do work (UNESCO, 2020a; Sudevan, 2020; Sahni,2020). On the whole, the COVID-19 pandemic put the education system in jeopardy. The challenge thus imposed forced educational institutes to adopt digital technology in learning.

### 8.5.1 Shifting Pedagogy and Student Assessment: From Face to Face to Online Learning

COVID-19 forced a change in learning from conventional educational models to educational technology (EdTech) models, resulting in a rapid transition from physical learning to digital learning (Dhawan, 2020). Online learning has been found to be the best possible alternative to conventional learning (Adnan and Anwar, 2020). It provided a chance to develop new and improved methods of online learning that are more useful and efficient.

During the pandemic lockdown, e-learning resources were critical in assisting educational institutions in facilitating student teaching (Subedi et al., 2020). There has been a paradigm shift in the way educators deliver quality education through various online platforms. Transitioning from traditional face-to-face learning to online learning can be an entirely different experience for learners and educators. On the direction of the Ministry of Education, Government of India, educational institutions started online classes on platforms like Google meet, Zoom, Web ex, Microsoft Teams, etc., and promoted digital learning through e-learning platforms like SHAGUN, DIKSHA, SWAYAM, and e-PATHSHALA. Online classes at secondary school level were available on 32 DTH channels on TV along with Tata Sky and Dish TV. These portals provided free study materials to all students. Even though schools were closed, students could download e-books from the websites and access learning modules.

Appraisal and assessment are other important aspects of educational learning affected by the COVID-19 lockdown. A crucial part of online distance learning is the availability of helpful formative assessments and timely feedback to online learners (Doucet et al., 2020). To prevent the spread of Covid, online examination or automatic promotion on the basis of previous results were introduced. Student assessments were carried out online, with a lot of trial and error, uncertainty, and confusion among teachers, students, and parents. Various approaches to assessment were used and tools to check plagiarism, etc., were put in place since student assignments and examinations were carried out from home. Moreover, many parents guided and supported their children during their learning process though the extent and degree of support varies greatly. Thus, objective and authentic assessment of students was found to be challenging for educators and institutions. It was even more challenging in some parts of India due to larger class sizes, lack of online teaching infrastructure and professional development, and the non-participative nature of some students.

### 8.5.2 Opportunities for Innovative Teaching and Learning

While there have been many obstacles for educators, colleges, institutes, and the government with regards to online education, the COVID-19 pandemic provided several opportunities to enhance e-learning.

*Online teaching* helped people learn to use digital technology and resulted in increased digital literacy and skills. Teachers upgraded their digital skills in an effort to guide students through online learning. Digital learning platforms provide students with full access to study material as well as allow them to engage in online classes and interact with teachers just like in a physical classroom setting. For the first time ever, online platforms such as Google Classroom, Zoom, interactive learning environments, social media, and various community channels such as Telegram, Messenger, WhatsApp, and WeChat were explored and used for teaching and learning. The pandemic also popularized online conferences, meetings, and webinars.

While there are advantages to online teaching, there are also limitations, and teachers are expected to to find ways to improve online teaching methods and address the challenges. But when educators, parents, and students share common experiences, there are unrivalled opportunities for collaboration (Doucet et al., 2020). Online teaching has provided the opportunity to teach and learn in innovative ways different from the teaching and learning experiences in the normal classroom setting. In addition to technological innovative pedagogy, it has strengthened the bond between teachers and parents. Parents have also had the opportunity to further bond with children by providing academic support for their children's education through homeschooling and online learning.

### 8.5.3 Challenges Faced by Teachers in Online Education

Since online learning is different from the traditional face-to-face teaching, teachers had to adapt to meet the changing needs of students in a short amount of time. Quickly adapting technology for online teaching was a major challenge for Indian teachers. Along with teaching online classes, teachers were also required to administer and evaluate online quizzes and exams. As noted by Punit and Qz.com (2020), online classes amid lockdown resulted in teachers facing challenges related to their institutions, technology, lack of training, and home environment and students' attitudes. A survey conducted by QS I GAUGE found that the higher educational institutes did not have the required technological infrastructure to ensure sound delivery of online classes to students (QS-ERA, 2020). The survey found that only a few institutions had been using advanced teaching apps like Moodle, online Blackboard, Microsoft Teams, and Google Meet for teaching but most used open-source platforms such as WhatsApp, YouTube, Skype, Google hangout, Zoom, etc., for online teaching.

Joshi, Vinay and Bhaskar (2020a) also observed that online teaching is affected by home environment on the one hand and institutional barriers on the other. Lack of proper training, lack of technical support, and lack of institutional support to provide basic infrastructural facility with clarity and direction were some of the various institutional factors affecting the quality of online teaching. Many institutions did not provide clear instructions and directions about the implementation and execution of online classes and expected all teachers to know how to do online classes on their own (Sharma, 2020). Further, Joshi, Vinayand Bhaskar (2020b) and Kaup, Jain, Shivalli, Pandey and Kaup (2020) also noted that many institutions did not have trained teachers able to work remotely through online platforms and struggled with the transition, lack of availability of configured laptops, internet, microphones, etc., to efficiently teach online. A number of teachers faced connectivity issues, system failure, and bandwidth issues while conducting online sessions and due to lack of technical assistance were often unable to resolve problems.

Online teaching was described by teachers as a draining and demotivating experience. A typical classroom has chairs, desks, a whiteboard, and a projector. Teachers were forced to teach at home without these tools and also deal with external disturbances such as traffic from the street or interruptions by family members during lecture (Press Trust of India, 2020).

Teachers were forced into online education, which resulted in hostile views toward institutions. Shenoy, Mahendra and Vijay (2020) reported that many teachers were not happy to adopt technology and virtual classrooms and were not comfortable teaching courses that required numerical calculations, tests, and interpersonal contact. Cultural barriers are reduced during face-to-face instruction when teachers use bilingual dialogue to answer students' doubts and questions. On a small screen, teachers found it difficult to assess students' body language, and emotional connection was lacking. The value of classroom learning cannot be overstated, and the pandemic demonstrated that educational institutions and teachers were not equipped to accommodate multimedia instruction (Mahesh, 2020; Azevedo, Hasan, Goldemberg, Iqbal & Geven, 2020).

In terms of home atmosphere, teachers reported family distraction and interference, a lack of independent space at home, and personal issues some hurdles in effective online education. Arora and Srinivasan (2020) added network challenges, a lack of knowledge, a lack of interest, low attendance, lack of personal touch, and a lack of interaction to the list of factors influencing online teaching. (Joshi, Vinay and Bhaskar, 2020b; Sudevan, 2020; Almaiah, Al-Khasawneh and Althunibat, 2020). Managing students in online classes presented numerous problems and difficulties for teachers. Teachers reported students deliberately by playing music, making noise, posting bad comments through fake user IDs, eating, and playing games, which further hindered online learning.

### 8.5.4 Challenges Faced by Students in Online Education

Online learning undermines social skills. Aside from being enjoyable for children, school time helps them develop social skills and awareness. When students are away from their regular school schedule, they face physical, social, and psychological consequences. Some students were also at risk of online exploitation and were exposed to potentially dangerous and abusive content, as well as a higher risk of cyber bullying, as a result of increased and unstructured time spent on online. School closures and strict containment measures meant more families were dependent on technology and digital solutions to keep children engaged in learning, entertained, and connected to the outside world, but not all children have the necessary knowledge, skills, and resources to keep themselves safe online.

Inherently driven learners were largely unaffected in their learning since they need less instruction and encouragement, while students who are poor in learning faced challenges. Academically good students belonging to lower socio-economic background may be unable to access or afford online learning due to lack of space at home. As a result, a number of students experienced social and emotional trauma, making them unable to attend online classes or communicate effectively. The best practices for online homeschooling are yet to be explored (Petrie, 2020).

Alvarez (2020) and Mishra (2020) discussed numerous concerns related to educational inequality caused by the digital divide and the lack of required infrastructure to upgrade online education with equality in India. Students also encountered significant difficulties in obtaining stable internet service and connections to digital devices. Economically disadvantaged children cannot afford multimedia learning equipment. Online schooling also puts learners at risk of excessive screen time. As a result, offline practices and self-exploratory learning have been essential for students. Another problem is a lack of parental supervision, particularly for young learners. There are practical issues around physical workspaces conducive to different ways of learning.

Lack of technology or adequate network access is a barrier to continuous learning, particularly for students from low-income families in rural and underprivileged communities (UNESCO, 2020a). In the absence of schools, children and youth are deprived of essential learning and opportunities for normal growth and development. The resulting disruptions worsened the already existing disparities within the education system and other aspects of their lives. Underprivileged learners belonging to marginalized communities were doubly disadvantaged, since they tend to have fewer educational opportunities beyond school. The education of a large number of rural and urban children suffered a lot due to lack of access to technology and internet connectivity. The NHRC (National Human Rights Commission) also remarked that with only 15% rural households having access to the internet, the education of over 250 million students has been adversely impacted. "There have already been multiple cases of student suicides due to the helplessness of not being able to access education digitally. An assessment by Oxfam found that 75% of government school parents struggled to support their children in accessing education that was delivered digitally." Students with special needs, such as hearing disability, vision impairment, or mobility disorders, need extra instruction, as well as counseling and guidance. Many guardians and parents at home are unable to meet such needs, which impedes the learning of these students. As a result, there is a need to devote more time and resources to investigate and research the best possible alternatives for the special educational needs (SEN) of these learners with regard to the future.

Children have also faced malnutrition due to lack of quality food intake. Owing to closure of schools, a large number of children studying in government schools were deprived of mid-day meals and thus became malnourished. The NHRC of India reported concern for food security of children because of the interruption in the supply of meals and supplementary nutrition under the Mid-Day Meal Scheme (MDMS) and Integrated Child Development Services (ICDS).

The Human Rights Commission issued an "Advisory for Protection of the Rights of Children in the context of COVID-19" and urged seven Union ministries and all state governments and union territories to take specific actions for promotion and protection of the rights of children under five in broad categories –health, food and nutrition, education, childcare institutions, and child protection and justice systems.

Many have experienced tension, anxiety, and depression as a result of COVID-19, thus future research should look at the effects of the pandemic on students' mental health. In addition, recommendations should be developed to ensure that mental health services are provided.

Reserachers should also closely examine the problems that arose as a result of the abrupt switch to online learning and be prepared for the future. Educators must be properly trained in digital skills and student-teacher engagement must be strengthened. To prevent any interruption to their studies, deprived students must have access to computers and internet connectivity.

## 8.6 CONCLUSION

COVID-19 has had a significant impact on family life and educational spheres in India. Many families faced a great deal of stress with the added strain of caring for young children and the elderly. Some vulnerable family members also experienced violence, tension, and uncertainty as a result of lockdowns. On the other hand, some families used lockdown periods to strengthen and reconnect with family members, rejuvenate family relations, pursue new activities, and to practice yoga and meditation. However, it is too early to tell whether these changes will become a normal and lasting feature of family life.

COVID-19 has had an immense impact on the educational system of India. Though it has posed many challenges, some opportunities have also evolved. Innovative educational methods were developed to impart education, appraisal and evaluation, admissions, and educational administration. Digital platforms ensured continuation of online classes, and provided a chance to develop new and improved professional skills/knowledge through online learning. However, both teachers and students encountered difficulties in adapting to technology-based educational options due to lack of technology and internet access, and limited online teaching and learning skills. More research is needed to understand the effects of COVID-19 and to develop new policies and recommendations to prepare for the future to ensure preparation, equality, and access.

## REFERENCES

Addati, L; Cattaneo, U; Esquivel, V and Valarino, I (2018) The Care Economy. International Labour Organisation (ILO) http://suo.im/6fzKfq.

Almaiah, M A; Al-Khasawneh, A and Althunibat, A (2020). Exploring the critical challenges and factors influencing the E-learning system usage during COVID-19 pandemic. *Education and Information Technologies, 25,* 5261–5280.

Alvarez, A J (2020). The phenomenon of learning at a distance through emergency remote teaching amidst the pandemic crisis. *Asian Journal of Distance Education,* 15(1), 127–143.

Arora, A K and Srinivasan, R (2020). Impact of pandemic COVID-19 on the teaching–learning process: A study of higher education teachers. *Prabandhan: Indian Journal of Management,* 13(4), 43–56.

Azarian, B. (2020). Family distancing: importance and psychological effects, Psychology Today, March 27. www.psychologytoday.com (accessed August 15, 2020).

Azevedo, J P; Hasan, A; Goldemberg, D; Iqbal, S A and Geven, K (2020). Simulating the potential impacts of covid-19 school closures on schooling and learning outcomes: A set of global estimates. The World Bank. Retrieved from https://reliefweb.int/report/world/simulatingpotential-impacts-covid-19-school-closures-schooling-and-learning-outcomes

Bloomberg (2020). China's divorce spike is a warning to rest of locked-down world. March 31. Retrieved from www.bloomberg.com/news/articles/2020-03-31/divorces-spike-in-china-after-coronavirus-quarantines

Buheji, M; Ahmed, Dunya and Jahrami, H (2020) Living uncertainty in the new normal, *International Journal of Applied Psychology,* 10(2), 21–31.

Dhawan, S (2020). Online learning: A panacea in the time of COVID-19 crises. *Journal of Educational Technology,* 49(1), 5–22. https://doi.org/10.1177/0047239520934018

Doucet, A; Netolicky, D; Timmers, K and Tuscano, F J (2020). *Thinking about pedagogy in an unfolding pandemic* (An independent report on approaches to distance learning during COVID-19 school closure). Work of Education International and UNECSO. https://issuu.com/educationinternational/docs/2020_research_covid-19_eng ..Google Scholar

Ghosh S (2015). The political economy of domestic violence in a Mumbai slum: An Ethnographic Analysis. *Journal of Interdisciplinary Economics*, 27(2), 1–24.

Joshi, A; Vinay, M and Bhaskar (2020a). Impact of Coronavirus Pandemic on the Indian education sector: Perspectives of teachers on online teaching and assessments. *Interactive Technology and Smart Education*. Retrieved from www.researchgate.net/publication/343961926_Impact_of_Coronavirus_Pandemic_on_the_Indian_Education_Sector_Perspectives_of_Teachers_on_online_teaching_and_a ssessme nts

Joshi, A; Vinay, M and Bhaskar, P (2020b). Online teaching amidst COVID-19 in India: An outlook. *Asian Journal of Distance Education,* 15(2), 105–111.

Kasarla Harshitha, 2020, The Wire

Kaup, S; Jain, R; Shivalli, S; Pandey, S; and Kaup, S (2020). Sustaining academics during COVID-19 pandemic: The role of online teaching-learning. *Indian Journal of Ophthalmology,* 68(6), 1220–1221.

Liu, Y (2020) As many countries around the world begin to emerge from lockdown, what can we learn from how people's relationships and friendships have fared in China, where coronavirus began? June 5. www.bbc.com/future/article/20200601-how-is-covid-19-is-affecting-relationships

Mahajan, S (2020).*Technological, social, pedagogical issues must be resolved for online teaching* [Web log post]. Retrieved from https://indianexpress.com/article/opinion/columns/indiacoronavirus-lockdown-online-education-learning-6383692/

Mahesh, S (2020). A need now but no replacement: Teachers share concerns about online classes during COVID-19 [Web log post]. Retrieved from www.newindianexpress.com/education/2020/may/06/a-need-now-but-no-replacementteachers-share-concerns-about-online-classes-during-covid-19-2139 605.html

Mehta, J C (2014). *Globalization, Economy and Society*. New Delhi: Concept Publishing Company.

Mehta, J C (2017). *Social Structure and Educational Inequalities*. New Delhi: Concept Publishing Company.

Mishra, S V (2020). COVID-19, online teaching, and deepening digital divide in India. Retrieved from /www.researchgate.net/publication/341889105_COVID-19_online_teaching_and_deepening_digital_divide_in_India

Mukhtar, S (2020). Psychological health during the coronavirus disease 2019 pandemic outbreak. *Int J Soc Psychiatry*, 66(5), 512–516. 10.1177/0020764020925835

Petrie, C (2020). *Spotlight: Quality education for all during COVID-19 crisis* (hundrED Research Report #01). United Nations. Retrieved from https://hundred.org/en/collections/quality-education-for-all-during-coro navirus Google Scholar

Power, K (2020) The COVID-19 pandemic has increased the care burden of women and families, Sustainability. *Science, Practice and Policy*, 16(1), 67–73. https://doi.org/10.1080/15487733.2020.1776561

Prasso, S (2020, March 31). Divorce Rate After Coronavirus Quarantine in China Is Warning. Retrieved from www.bloomberg.com/news/articles/2020-03-31/divorces-spike-in-china-after-coronavirus-quarantines

Pravat, K J (2020a). Challenges and Opportunities created by Covid-19 for ODL: A case study of IGNOU. *International Journal for Innovative Research in Multidisciplinary Filed*, 6(5), 217–222.

Press Trust of India. (2020, April 9). From technological queries to distress calls, teachers struggle with challenges posed by lockdown [Web log post]. Retrieved March 10, 2021 from www.ndtv.com/educat ion/from-technological-queries-to-distress-calls-teachersstruggle-with-challenges-posed-by-lockdown-2208957

Punit, I S and Qz.com. (2020, May 13). For many of India's teachers, online classes amid lockdown have been an awful experience [Web log post]. Retrieved from https://scroll.in/article/961738/formany-of-indias-teachers-online-classes-amid-lockdown-have-been-an-awful-experience

QS-ERA (2020). *COVID-19: A wake-up call for Indian;* QS-ERA India Pvt Ltd.

Ravichandran, P and Shah, A K (2020 July). Shadow pandemic: Domestic violence and child abuse during the COVID-19 lockdown in India. *International Journal of Research in Medical Sciences*, 8(8), 3118. https://doi.org/10.18203/2320-6012.ijrms20203477 Google Scholar

Roshgadol, J (2020) Quarantine quality time: 4 in 5 parents say coronavirus lockdown has brought family closer together. Study Finds. http://suo.im/6uxYkC.

Rukmini.S (2020). Locked down with abusers: India sees surge in domestic violence. Retrieved from www.aljazeera.com/news/2020/04/locked-abusers-india-domesticviolence-

Sahni, U (2020). COVID-19 in India: Education disrupted and lessons learned [Web log post]. Retrieved from www.brookings.edu/blog/education-plus-development/2020/05/14/covid-19-inindia-education-disrupted-and-lessons-learned/

Saji J A; Babu B P and Sebastian S R (2021). Social influence of COVID-19: An observational study on the social impact of post-COVID-19 lockdown on everyday life in Kerala from a community perspective. *Journal of Education and Health Promotion*, 9, 360.

Sharma, K (2020). Why online classes may not be such a good idea after all, especially for kids [Web log post]. Retrieved from https://theprint.in/india/education/why-online-classes-may-not-besuch-a-good-idea-after-all-especially-for-kids/406979/

Shenoy, M V; Mahendra, M S and Vijay, M N (2020). COVID 19–Lockdown: Technology adaption, teaching, learning, students engagement and faculty experience. *MuktShabd Journal*, 9(5), 698–702.

Subedi, S; Nayaju, S; Subedi, S; Shah, S K and Shah, J M (2020). Impact of e-learning during COVID-19 pandemic among nurshing students and teachers of Nepal. *International Journal of Science and Healthcare Research*, 5(3), 9.

Sudevan, P (2020). Why E-Learning isn't a sustainable solution to the COVID-19 education crisis in India [Web log post]. Retrieved from www.thehindu.com/sci-tech/technology/whyelearning-is-not-a-sustainable-solution-to-the-covid19-education-crisis-inindia/article31560007.ece

Szabo, T G; Richling, S; Embry, D D et al. (2020) From helpless to hero: Promoting values-based behavior and positive family interaction in the midst of COVID-19. *Behaviour Analysis in Practice*. https://doi.org/10.1007/s40617-020-00431-0

The Times (2020, March 18). As cities around the world go on lockdown, victims of domestic violence look for a way out. *Time*. Retrieved from https://time.com/5803887/coronavirus-domestic-violence-victims

UN (2020). Put women and girls at centre of efforts recover of COVID-19, Secretary-General, United Nations. http://suo.im/5wohwt

UNESCO (2020a). Adverse consequences of school closures, UNESCO. Retrieved from https://en.unesco

UNESCO (2020b). *290 Million students out of school due to COVID-19*. UNESCO releases first global.

Woodruff, A (2020) Keeping the family in family medicine. *American Journal of Hospice and Palliative Medicine*, 1–2.

# 9 COVID-19 Pandemic's Impact on Education in India

## Reimagining Education beyond the Pandemic

*Satvinderpal Kaur*

Department of Education, Panjab University, Chandigarh, India

## CONTENTS

## 9.1 THE CONTEXT

The COVID-19 pandemic has had wide-ranging and intersecting impacts on the economy, health system, educational system, and ways of life globally. The pandemic has not only become a global crisis, but also a sort of prism that has exposed and magnified the underlying issues of contemporary society and prompted us to think about our previous assumptions of development. COVID-19 caused the largest ever closures of educational institutions, spanning over 180 countries, implemented to curb the virulent spread of the virus. It has been estimated that around 1.6 billion children and youth were out of school during this time (World Bank, 2020). By January 2021, almost 825 million students were affected due to the closure of educational institutions amid pandemic. According to UNICEF (2021) monitoring report, 23 countries have implemented the nationwide closures and 40 are implementing local closures, impacting about 47% of the world's student population. In South Asia, disruption to schooling and the accompanying economic dislocation was widespread and prolonged, with severe consequences for poor, rural populations, and especially for women and girls. The United Nations also estimated that nearly 600 million children in South Asia were affected by lockdowns imposed in March or April (Menon, 2020).

In India, education took a massive hit due to the pandemic and the resultant lockdowns beginning from March 2020. All of a sudden, the wheel of moving life came to halt in an unprecedented manner. According to UNICEF, the COVID-19 pandemic has battered education systems. In India, over 1.5 million schools closed down due to the pandemic, affecting 286 million children from pre-primary to secondary levels. This adds to the 6 million girls and boys who were already out of school prior to COVID-19. This crisis is affecting most heavily vulnerable populations including children, women, people with disabilities, rural and tribal groups, and the poorest households. It is reported that the loss of education caused by school closures can impact the health and nutrition

DOI: 10.1201/9781003358916-9

of children in the long term. School meals are a vital source of nutrition among schoolchildren. Since the beginning of the pandemic, 357 million school children have not been receiving school meals in 150 countries across the globe. More than 39 billion school meals were missed globally due to school closures during the pandemic. In India the mid-day meal scheme has been shown to decrease calories by 30% and nearly 50% of the children are already malnourished (UNICEF, WP, January, 2021).

The COVID-19 pandemic has made visible the economic, health, caste-based, gender, and educational inequalities the underprivileged face in India. In the absence of real educational sites, remote learning has been established as an alternative while all the educational institutions remain closed. Unfortunately, previous policy paralysis has inhibited the government from reimagining post-pandemic education, despite the loss of livelihood, food and shelter brought by educational inequalities and institutionalised by the neoliberal reforms (Batra, 2021). The pandemic has forced nations such as India to re-examine the meaning and purpose of education from a socio-cultural perspective. To explore the impact of COVID-19 on Indian education systems policy makers must consider the context of existing realities by scrutinising the assumptions manufactured by the market agenda of education.

## 9.2   ONLINE EDUCATION AND THE STARK SOCIAL INEQUALITIES

Calamities, be they natural or manmade, affect the underprivileged the hardest, and COVID-19 is no exception. The pandemic has revealed the rooted structural inequalities between rural and urban, rich and poor, male and female even in the digital world. The rhetoric of development has been exposed with the figures indicating that the richest 1% of the population has four times the wealth of the bottom 70% in India, in large part due to takeover and monetisation of the land, forests, mountains, and other natural resources, which were the livelihood of those at the bottom rung of the social hierarchy (Chaudhry and Aga, 2020b). The sudden draconian lockdowns during the pandemic have also uncovered the delicate existence of hundreds of millions of Indian people and the fate of their children who sustain just to meet both ends.

Ever since the outbreak of the pandemic, the pedagogical challenges have been multiplied. Online learning has become a critical lifeline for education as classroom teaching has shifted to digital platforms. The government also encouraged online learning. India's National Commission for UNESCO (2020b) issued a short paper detailing support for distance learning during lockdown focusing primarily on resources for digital learning (INCCU, 2020). However, in the absence of physical classes and adequate digital infrastructures both students and teachers are facing a great challenge. Despite the efforts made by the government to deploy online and other distance learning tools, the World Bank accepted that given the country's large digital divide, technological solutions alone would have limited impact to fill the vacuum of real classroom losses. The Bank observed that disadvantaged students are likely to be the most impacted and that poorer children will likely to fall behind their wealthier counterparts who have better access to online learning resources (Gupta and Khairina, 2020). Hence, the children from poor, lower middle-class, and migrant worker families who lie at the other side of digital divide have not been able to have access to any form of online education. Clearly in the new normal of teaching through online mode, the already existing digital divide gets more sharpened.

The socio-economic inequalities have a bearing on learning opportunities and access to online resources. According to a National Sample Survey (2017–18), the key Indicators of Household Social Consumption on Education in India reported that less than 15% of rural Indian households have internet access, as opposed to 42% in urban Indian households. A mere 13% of people surveyed (aged above five) in rural areas and a mere 8.5% of females could use the internet. The poorest households cannot afford smartphone or computers. The digital divide is apparently seen across, gender, class, and place of residence. Amongst the poorest 20% households, only 2.7% have access to a computer and 8.9% to the internet, while in the top 20% of households, this proportion is 27.6% and 50.5%, respectively. Also, the government figures indicate that fewer than a quarter of Indian households have internet access, although radio and television are accessible to many (Menon, 2020). A significant gender divide in access to technology is seen, with women's access to mobile phones often governed by male relatives being just one factor threatening a serious reversal in gender equality as a result of school closures (Nikore, 2020). To even the NEP (2020), aimed at increasing online learning in schools and at higher education level without bridging social and economic inequalities and by ignoring the digital divide, more so in the tribal and remote areas across the nation. Many reports found that the challenges of remote learning are disparities in access, ranging from electricity and internet connections to availability of devices like computers and smartphones. Not only this, learning needs a conducive, peaceful, and undisturbed environment. As per Census 2011, 37% of the Indian households have only one dwelling room. Stable internet access comes after the establishment of an able social infrastructure, which demands more public sector investment in infrastructure, but unfortunately both are negligible in developing countries like India. In the last three decades, the policies of development ushered in by the neo-liberal agenda resulted in asymmetrical development in India. And the unequal distribution of wealth and resources has sharpened since the dawn of neoliberalism. Like other resources, the digital divide has further worsened the polarisation between the haves and have-nots.

While the long-term implications of disruption to schooling as a consequence of COVID-19 remain unclear, the significant effects on child welfare and inequality, given uneven internet access and other differences in family resources for supporting learning at home, are clear. A significant reorientation of education policy, already in the pipeline before COVID-19 struck, was promulgated in the summer of 2020. This policy combines the enhancement of state control with a tighter embrace of marketisation and technology by increasing the use of technology in education and expanding online education and training for teachers. Due to the existing digital divide expanding online learning will only further push the digital havenots to the periphery of the education system.

## 9.3 LEARNING LOSSES AND EDUCATIONAL EXCLUSION

The COVID-19 crisis will further compound learning poverty and reduce progress in meeting the SDG-4 commitments relating to education and to ensuring inclusion and equity. It is extrapolated from the existing data that the global level of education will fall, and up to 7 million children could drop out of school due to the economic impacts of the COVID-19 (World Bank, 2020). For India, the constitutional imperative carries a far more empowering framework for educational and other social rights than the SDG-4. There is no doubt that the school closures in the first half of 2020 resulted in significant learning losses to the affected cohort of students. Many studies across the world have clearly indicated that school closures have significant negative impacts on learning levels of children and children from disadvantaged backgrounds are affected more severely. The multiple deprivations and digital divide will further exacerbate the already incurred learning losses. It is also estimated that these learning losses will further push students into the labour market, and as a result both students and their nations will likely to confront the adverse economic outcomes. The economic losses will be more deeply felt by disadvantaged students as envisioned by UNESCO 2020a. All indications point out that students whose families are less able to support out-of-school learning will face more learning losses than their advantaged peers and that ultimately will translate into deeper losses of lifetime earnings.

A survey conducted by the NGOs that work with children, organised by Kailash Satyarthi, found that 85% of the responding organisations felt that school drop-outs are likely to increase in the post-lockdown period In India. With losses in household income, expectations that children contribute financially can intensify. It is reported that out of these children, more girls could be forced into seeking jobs. Children who were already working may do so for longer hours or under worsening conditions. Gender inequalities may also grow more acute within families, with girls expected to perform additional household chores and agricultural work rather than to attend school.

This loss of learning is not simply the curricular learning that children would have acquired if schools remained open, rather, it includes the abilities that children have forgotten due to lack of usage, such as the ability to read with understanding, the ability to write, and the ability to perform basic mathematical operations like addition and multiplication. This regression further compromises new learning since these abilities are foundational to all further learning. This situation must be juxtaposed with the fact that we are already facing a crisis in learning, particularly with respect to foundational literacy and numeracy. The World Bank (2020) noted that even before the pandemic, the world was facing a 'learning crisis'... 258 million children and youth at primary and secondary school level were out of school and low education quality means those who were in school, learn too little. The new push for online learning will also initiate critical debates on the learning crisis. OECD (2020) reported on the impact of learning loss in economic terms and showed that across the world the economic impact of learning losses will be huge in relation to education spending of GDP. These learning losses will be the worst for socio-economically disadvantaged students as household income and family environment are the major determinants of children's academic achievement in normal circumstances. Socio-economically advantaged parents also tend to compensate for any deterioration in schooling to a greater extent, as compared to disadvantaged parents. The learning losses are difficult to quantify, since the qualitative losses of missing learning spaces might be far beyond the quantitative estimations, as the prolonged lockdowns and uncertainty cause lifetime socio-psychological disorders. Moving students to online has starkly exposed the deep inequities in education systems, because a huge number of children who rely on school for food and safe environment are deprived of both. According to Vickers (2020) access to schooling is vital not only to the welfare of children and their families, but also to any veneer of credibility. If children are denied instruction, then even the appearance of equality of opportunity is fatally undermined. Hence to exercise the experience of education with the physical presence is their basic right. The denial of which is against their right to education.

## 9.4   CHALLENGE TO THE IDEA OF EDUCATION

Scrutinising the role of education in the current pandemic and socio-ecological crisis is critical to understanding the true idea of education. The dominant international discourse on education continues to see education primarily as a tool to enhance economic growth. It compels us to locate its roots in an educational policy shift from public to market. Though intellectuals have continued to assert against replacing the humanistic approach to education by the mechanised and instrumental model, the market model of education ushered in by neo-liberal policies has side-lined the true meaning and purpose of education. The aims of education envisioned by Gandhi and Tagore to self-realisation and perfect preservation of the symphony of knowledge between one's life and outside world has been replaced by mere competencies and skills required for the labour market. In the words of Cunningham (2020) it is human behaviour that causes conditions of destruction on nature, which creates an opportunity for pathogens like COVID-19 to grow on other hosts. The current model of education works against the sustainability of the world and environment. The real purpose of education is to advance and disseminate knowledge and this is undermined by the logic of the market to maximise private benefits. As described by Noam Chomsky, educational institutions are becoming anti-educational sites, not only as a result of greater competition, but due to the relentless focus on tests and achievements, which have diminished the crucial potentialities of individuals. The educational thought processes are determined by companies for private profit, and learning ideas are reduced to mechanistic ritual only, rather than an organic process freely engaged in to maximise the range of understanding. During the pandemic online education is seen as an alternative to physical classrooms but big corporate interests are lying behind this shift.

With the undue focus on instrumental aims of education, knowledge itself is being marginalised. On this changing purpose of education and knowledge, Kumar (2020) writes that the knowledge brought to life through technology is irrelevant and is its value as truth. With the loss of pedagogical physical spaces and adoption of online modes of teaching, the very idea of education has been distorted. The idea that learning can be reduced to the individualised transmission of skills, all with the aim of maximising 'human capital', represents the apotheosis of the machine and the collapse of a humane conception of education. The World Bank's strategy since 1994 has been based on the promotion of a model of knowledge adjusted to the requirements of market and a market model of education delivery that involves privatisation, commercialisation, and corporatisation of education. Under this model the teaching learning processes are reduced to acquiring skills and procedural competencies rather than knowledge that develops critical thinking, creativity, and conceptualisation of the life realities essential for a holistically developed citizen (Prasad, 2020).

As educational institutions have remained inaccessible for more than two years, this crisis has been seen as an opportunity to enhance a neo-liberal focus on learning outcomes via online learning support systems, with little acknowledgement of the fact that a larger proportion of the population does not have internet access (Batra, 2021). The pandemic has exposed the market model in self-explanatory terms. The prolonged lockdowns brutally exposed the malign consequences of the policies of privatisation in India by looking at low-cost private schooling as an alternative to government schools. In early October 2020, the *Times of India* reported that, in the Indian state of Telangana alone, close to 10,000 private budget schools were staring at closure and approached the state government for a takeover (Biswas, 2020). Despite the COVID-induced crisis in low-cost private schooling, the NEP 2020 envisaged further involvement of private or 'philanthropic' actors in educational provisions, calling for state schools to be paired with private counterparts to enable sharing of 'best practices' (Prasad, 2020). Unfortunately, the government has failed to commit to a common school system based on neighbourhood schools for all children, irrespective of socio-economic status (Sadgopal, 2020). Clearly, the neglect of public services like health and education cause Indian society to suffer and have posed a big challenge particularly during the pandemic.

The pandemic has put tremendous psychological and mental stress on children. COVID-19 dramatised this fact for millions of learners, isolated from each other and from their teachers, separately accessing information online. An instrumentalist vision of education-as-human-capital-generation, allied to a narrowly technocratic outlook, diminishes or dismisses values intrinsic to a fulfilling life. One should not forget that education cannot thrive with ready-made content built outside the pedagogical space and human relationships between teachers and students. The policies and plans to design and measure education based on mere learning outcomes is a big mistake. As only quality becomes a matter of judging the outcomes of education, and this outcome-driven strategy of judgement and evaluation hits the poorest sections the hardest (Kumar, 2020). In India, education is supposed to teach inner resilience and a sense of social justice, but the dominant international and national education discourses based on the human capital approach have continued to view education from the perspective of employability and economic growth. This is evident from the decision of lockdown and closure of educational institutions and shifts to online teaching despite the widespread asymmetric distribution of resources and unequal access to technology. By not focusing on skills aimed at sustainable development, the market model of education is producing educated individuals who may not only be psychologically equipped to deal with challenges like COVID-19 but also insensitive to the sufferings of others. The COVID-19 crisis can serve as an opportunity for India to redefine and reimagine its education system.

## 9.5    REIMAGINING EDUCATION BEYOND THE PANDEMIC

The profound alienation and social fragmentation have posed big challenges to millions of students. It is time to reimagine education to ensure learning continues with minimal disruption but that also empower students as compassionate human beings prepared for an unpredictable future and citizens seeking a peaceful and kind world. This pandemic forced us to rethink how education works to equip students with the cognitive, creative, social, emotional, and physical skills required to prepare them for life and not merely for livelihood. Our hopes for a peaceful, prosperous and environmentally sustainable future depend on grasping the broader meaning and potential of education (UNESCO-MGIEP 2017). Education and curricula must be designed to help students understand the real-life challenges their community is facing. These include not only the climate change, but structural issues like socio-economic inequality, gender bias, and peace-building. Rethinking education is not enough; we must rethink the socio-economic models that education serves, moving away from a winners-take-all competition (Apple, 2014). A humane approach to education will remain elusive so long as we persist in ignoring the malign potential of education both as a vehicle for inculcating chauvinism and as a cog in the machinery of competitive meritocracy. Eliminating competition from our education system may be not possible, but we must recognise the dehumanising consequences of meritocratic fundamentalism. In the words of Vickers (2020), the purpose of education needs to change from being instrumental to human flourishing, accessible to all and structured in such a way where learning can happen anytime and anywhere and always. Many education approaches continue to focus mainly on rote memorisation, aiming to build the intellectual capacities of learners and attempting to make them job-ready – recognising high academic calibre with grades based on performance in standardised assessments. To focus education on the goal of increasing human capital is not sustainable. There is a manufactured scarcity of educational achievement possibilities. What we need education to focus on is the development of a more sustainable and peaceful way of living – with content, pedagogy, and assessments that facilitate human flourishing, constituting overall well-being and humanity. Policy makers urgently need make promoting peace, sustainability and a consciousness of shared humanity central to the visions for educational development.

## 9.6    CONCLUSION

The present health crisis has revealed our failures both at the social and political front. This crisis compels us to rethink our policies of development and our assumptions of education. Interventions are needed to support socio-economically disadvantaged students and those who have fallen behind academically. The marketisation of education ushered by promoters of neo-liberal policies and the values of efficiency, productivity, and output has promoted a culture of Taylorism in education by neglecting its true intrinsic value. The commercial interests in education have destroyed the critical thinking and creative-liberal space for learning. India needs to chalk out strategies on how current education systems may be reimagined and redesigned to equip future generations to develop a world that is more sustainable, peaceful, and future-ready. Policymakers must reflect on structural realities and the role of education and the potential changes it could make to nurture kind, inspiring, resourceful, and empathic learners. Disasters always teach lessons; the experience of going through a crisis creates the desire to rethink and reimagine shortcomings in existing systems. A nurturing of the holistic perspective of education needs to be looked into.

## REFERENCES

Apple, M. 2014. *Official Knowledge, Democratic Education in a Conservative age* (3rd ed.). New York, Routledge.
Batra, P. 2021. Reimagining curriculum in India: Charting the path beyond the pandemic. *Prospects,* 51 (407–424), 1–18.

Biswas, P. 2020. Telangana: Strapped private budget schools face closure. *Times of India.* https://timesofindia. indiatimes.com (accessed October 21, 2020)

Census 2011. Registrara Census, Government of India.

Chaudhary, C. and Aga, A. 2020. India's Pandemic response is a caste Atrocity. *NDTV opinion.* www.ndtv.com/ opinion/India-s-pandemic-response-is-a caste-atrocity-2236094. (accessed May 15, 2021)

Gupta, D. and Khairina, N.N. 2020. COVID-19 and learning inequities in Indonesia: Four ways to bridge the gap. *World Bank Blogs,* August 21. www.blogs.worldbank.org (accessed October 30, 2020)

INCCU. 2020. *Response to COVID-19.* New Delhi: Ministry for Human Resource Development (Indian National Commission for Cooperation with UNESCO). www.mhrd.gov.in/sites/upload_files/mhrd/files/ inccu_0.pdf (accessed October 21, 2020)

Kumar, K. 2020. NEP 2020 offers more of the same remedy. *Frontline Magazine.* Chennai: The Hindu Group. August 28, 9–13. Available at: https://frontline.thehindu.com/cover-story/it-offers-more-of-the-same-remedy/article32305017.ece (accessed April 12, 2021)

Menon, S. 2020. Coronavirus: How the lockdown has changed schooling in South Asia. *BBC Reality Check.* Delhi: BBC, September 21. www.bbc.com/news/world-south-asia-54009306 (accessed October 21, 2020)

Nikore, M. 2020. Covid Classrooms: India could lose progress made on girls' education. *Forbes India Blog,* July 13. www.forbesindia.com/blog/gender-parity (accessed October 21, 2020)

OECD. Data. 2020 Income Inequality. Paris: OECD. Available at: https://data.oecd.org/inequality/income-inequality.htm (accessed October 20, 2020)

Prasad, M. 2020. Education At the Mercy of the Market. *Frontline Magazine.* Chennai: The Hindu Group. August 28, 4–8.

Sadgopal, A. 2020. Decoding the Agenda. *Frontline Magazine.* Chennai: The Hindu Group. August 28, 9–13.

UNESCO Institute for Statistics. 2017. *More than one-half of children and adolescents are not learning world-wide.* Fact sheet 46. Montreal: UIS. http://uis.unesco.org/sites/default/files/ documents/fs46-more-than-half-children-not-learning-en-2017.pdf. read.ecd-ilibrary.org (accessed May 12, 2021)

UNESCO-MGIEP. 2017. Rethinking Education for the 21st Century: The State of Education for Peace, Sustainable Development and Global Citizenship in Asia. New Delhi: UNESCO-MGIEP.

UNESCO. 2020a. *Education in a post-COVID World: Nine Ideas for Public Action.* Paris: UNESCO International Commission on the Futures of Education.

UNESCO. 2020b. *Learning Never Stops. COVID-19 Education Response.* Paris: UNESCO. https://en.unesco. org/covid19/ educationresponse/globalcoalition. (accessed May 11, 2021)

UNICEF. 2021. Missing More than a Classroom, The Impact of school Closure on Children Nutrition, working paper 20–21.

Vickers, A. 2020. 'Covid-19, education and the Asian context: Meritocracy, inequality and Solidarity'. in *International Understanding and Cooperation in Education in the Post-Corona World, Asia-Pacific Centre of Education for International Understanding,* Edited by: Kwang Hyun Kim, JeongyeonSeo, Maggie Yang.

World Bank. 2020. The World Bank Group's response to COVID-19 pandemic. Report.

# 10 Telecommunication Interventions and Opportunities in Education, Health, and Agriculture
## Post COVID-19 Scenario

*Nitish Mahajan, Amita Chauhan, Harish Kumar,\* and Sakshi Kaushal*

UIET, Panjab University, Chandigarh, India
*harishk@pu.ac.in

## CONTENTS

## 10.1 INTRODUCTION

In the era of the Internet, advancements in technology have revolutionized the way people work and live, thus improving the quality of life. One of the biggest impacts technology has made is in the realm of telecommunications. With the evolution of mobile communication networks from GSM to current 4G LTE and forthcoming 5G, significant enhancements in data speeds and other services have been witnessed. Massive movement towards remote working has taken place, especially during the COVID-19 pandemic, and the telecommunications industry has made it efficient and effortless.

The key technology behind communication in 4G LTE and upcoming 5G is Voice over Internet Protocol (VoIP). VoIP is the transmission of voice and multimedia content over the Internet Protocol networks. The media is transmitted in the form of digital packets through packet switching instead of using circuit switching as in traditional networks like PSTN. The signaling steps involved in originating VoIP calls are similar to that of conventional networking methods; however, the analog streams are digitized and their packetization is performed in VoIP networks (Tagg et al., 2015). A special media delivery protocol is used for encoding video and audio with video codecs and audio codecs. Figure 10.1 shows the basic architecture of VoIP.

DOI: 10.1201/9781003358916-10

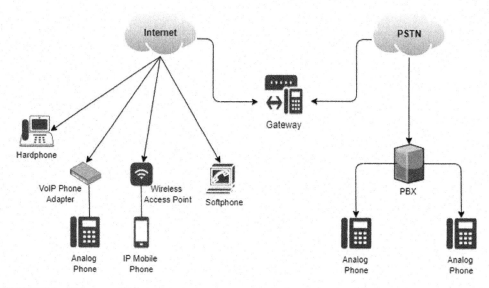

**FIGURE 10.1**   Voice over IP.

A brief description of VoIP components is given as follows:

1. **VoIP Clients:** Clients are the devices that interact with servers to get their calls routed. The two types of VoIP clients are hardphones and softphones. A hardphone is a conventional telephone look-alike that provides real-time audio communication. It is plugged into an IP network rather than a telephone line. It is basically a computing device that is configured through web services. A softphone is the software that is to be installed onto a computer, which allows it to provide real-time communication without using any dedicated hardware. The accounts are configured within the VoIP server by setting up extensions along with the username and password. The clients connect to these accounts within the VoIP server. In conventional PBXs, everything was configured within the PBX itself but for VoIP communication, the client has to provide the necessary information in the form of username and password to initiate the client-server interaction. If the username or password is incorrect, the VoIP server does not allow the client to get an extension.
2. **VoIP Server (IP PBX):** The VoIP server is the main server that routes calls through voice-over IP phones. It is also termed as IP Private Branch Exchange (IP PBX). The typical IP PBX can also switch calls between a VoIP user and a traditional telephone user, or between two traditional telephone users in the same way as a conventional PBX does. With a conventional PBX, separate networks are necessary for voice and data communications. The main advantage of an IP PBX is the fact that it employs converged data and voice networks. This means that Internet access, as well as VoIP communications and traditional telephone communications, are all possible using a single line to each user. This provides flexibility as an enterprise grows, and can also reduce long-term operation and maintenance costs. The IP PBX has a directory of all users and their corresponding addresses and hence can route an external call via VoIP gateway to the desired destination (Karapantazis et al., 2009).

In VoIP, call servers are also known as SoftSWITCH, which perform IP phone registration and provide call signaling. While IP PBX implements the VoIP features through hardware and software both, SoftSWITCH is a PBX-based system that relies entirely on software to provide VoIP services. It is basically a computer software with its source code made available with a license in which the copyright holder provides the right to study, change, and distribute the software to anyone for any purpose. SoftSWITCH is used to divide the call control function from media gateway and comprehends the call control and other functions like access control, interpretation, routing-selection, gateway management, call control, wide-band management, signaling, security, etc. There are different open source SoftSWITCHes available such as FreeSWITCH, YATE, Kamilio, Asterisk, etc. (Kaur et al., 2019).

The remainder of the chapter is organized as follows: Section 10.2 provides a survey of the concepts and features of the telecommunication network, Section 10.3 discusses the opportunities that telecommunications can bring to various application fields, and Section 10.4 provides a summary of the whole chapter.

## 10.2 LITERATURE SURVEY

### 10.2.1 TELECOMMUNICATIONS TECHNOLOGIES

Wireless communication technology using mobile devices has progressed over several decades. Since the 1980s, there has been advancement in the generations of telecommunications to the ongoing 4G and forthcoming 5G. In this section a brief introduction to the origin and advancements of the different generations of the wireless telecommunication technologies is presented (Farooq et al., 2013). Figure 10.2 shows how telecommunications connects the world at present.

**First Generation (1G)** The first generation of wireless telecommunication was 1G, which was used by most people for a long time. The maximum speed offered by 1G was 2.4Kbps with a bandwidth of 30KHz. Although 1G was the first wireless communication technology, data transmissions used to occur in analog form at a frequency of 150MHz. It is the "voice only" technology in which the only medium for conveying your messages was a phone call. Major drawbacks of 1G include poor voice quality, larger phone size, poor battery life, poor reliability, limited capacity, and insecure voice transmissions. To overcome these issues, 2G was launched after 10 years of 1G (Bohlin et al., 2010).

**Second Generation (2G)** The fundamental feature provided by 2G technology was digital transmission of signals. It offered services to convey content and picture messages at low speed and the bandwidth provided was 30 to 200KHz. The development of 2G technologies gave us a

**FIGURE 10.2** Telecommunication.

more secure connection, higher capacity, smaller devices, and better call quality as we moved from analog to digital telecommunication. The key advancements imparted by 2G over 1G technology were encrypted phone conversations, internet data services and Short Text Messages (SMS), and higher penetration levels for sending and receiving signals. GSM, GPRS, and EDGE are the fundamental technologies that come under the 2G spectrum. GDM stands for Global System for Mobile Communication. It is a cellular technology used to transmit mobile voice and data services. It is the most extensively accepted standard in telecommunications, and is administered globally. GSM is based on circuit-switched transmission and utilizes narrowband Time Division Multiple Access (TDMA) technique for transmitting signals (Samukic et al. 1998).

GSM provides a bit rate of 270Kbps within uplink frequency range of 933–960MHz and downlink frequency range of 890–915MHz. The three basic types of services offered by GSM are teleservices, bearer services and supplementary services. Teleservices provided by GSM to its users are voice calls, videotext and facsimile, and SMS. To transport data, teleservices use abilities of bearer services. In addition to teleservices and bearer services, GSM offers supplementary services like caller identification, call forwarding, conferencing, call waiting, call hold, call barring, number identification, etc. GPRS, short for General Packet Radio Service, was employed on the similar 2G technology as GSM, but with a few improvements to provide higher data speeds and easy billings. Enhanced Data rates for GSM Evolution (EDGE) had significant enhancements as compared to GSM and GPRS, and thus, was known to be 2.9G. EDGE provided a speed of 135Kbps and is still utilized on a number of mobile networks as it fulfills the fundamental needs of both users and carriers. 2G required powerful digital signals so as to make mobile phones work. Digital signals get weakened if there is no network coverage. Also, systems using 2G networks were unable to handle complex data such as videos. Hence, to overcome these limitations, 3G technology was developed (Patil et al., 2012).

**Third Generation (3G)** Deployment of 3G was a big revolution in respect of technological advancement for telecommunication. 3G offered faster data trans- mission speeds of up to 2Mbps and hence facilitated smart mobile phones to provide fast web browsing, video streaming, and the like. Maximum speed for non-moving devices as supported by 3G is around 2Mbps and 384Kbps for moving devices. 3G further evolved into 3.5G and then 3.75G as more features were added to the technology. 3G, unlike 2G, is based on packet switching data transfer. In addition to this, 3G allows its users to pay only for the actual data used rather than the time spent online. 3G has better encrypting techniques than those used in 2G, which provides better data security (Holma et al., 2005).

**Fourth Generation (4G)** With the evolution of communication technology and smartphones, the speed facilitated by 3G became insufficient. The next generation, called 4G, dramatically improved data transmission speeds. 4G supports mobile web access as well as provides services that require higher speeds such as video conferencing, gaming, HD mobile TV, 3D TV, etc. It offers a maximum speed of 100Mbps for mobile devices and up to 1Gbps for stationary or slow moving devices. 4G LTE and WiMax are the two technologies that were proposed under 4G. LTE stands for Long-Term Evolution, and it was the way out for achieving 4G speeds and standards. LTE was the simplification and redesign made in 3G architecture that resulted in remarkable reduction in signal transmission latency and increase in the network speeds (Ibrahim et al., 2002).

4G Voice Over LTE (VOLTE) is an advancement in LTE that completely digitizes voice telephony using packet switching approach, and provides enhanced voice quality along with better communication security. However, 4G technology has increased congestion in the communication network. Another key issue with 4G is that it is reaching the technological limits of how much data it can transfer quickly. The forthcoming 5G technology is thus the next generation of wireless communication (Akyildiz et al., 2010).

**Fifth Generation (5G)** 5G is envisioned to provide exponential increases in the transmission speeds over 4G LTE. The biggest difference between 4G and 5G is that 5G will act as a gateway for the IoT-connected sphere. Because of the new frequencies, spectrum, and technologies, 5G is

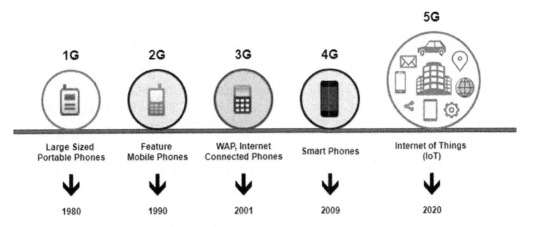

**FIGURE 10.3**   Generations of networks.

expected to offer better reliability, more security, less latency, higher speeds, less interference, larger capacity for devices, and better efficiency (Gohil et al., 2013).

Contrary to 4G LTE, 5G works on three different spectrum bands, namely low-band, mid-band, and high-band spectrum. Low bands of 5G, also called sub-1GHz spectrum, support blanket coverage and serve customers across urban, sub-urban, and rural areas. Low bands also provide IoT services, and make it simpler for the wireless signal to penetrate windows and walls (Mousa et al., 212).

To offer a good mixture of capacity and coverage, 5G has a mid-band spectrum. Spectrum in the range of 1GHz to 6GHz frequencies is mid-band spectrum. GSMA considers 3.3GHz to 3.8GHz range as ideal for 5G communication as it can carry a large amount of data and also travel significant distances. High bands are required to achieve the ultra-high speeds of broadband as envisioned for 5G. High bands are in the millimeter wave spectrum, and GSMA proposes that operators support 26GHz, 40GHz, 50GHz, and 66GHz frequency bands for mobile services as they have the strong momentum and international support (Maeder et al., 2016).

Architecture for 5G is designed to provide three key service categories, namely Enhanced Mobile Broadband (eMBB), Ultra Reliable Low Latency Communications (URLLC), and Massive Machine Type Communications (mMTC). eMBB initially is considered to be a natural evolution to currently existing 4G networks to provide faster data rates and better user experience. URLLC is architectured for critical applications like robotic surgery and autonomous driving that require high reliability and low latency. Loss of data packets or delay in the packet delivery for these kinds of applications during communication is not acceptable. The standard for URLLC provides error rates lower than 1 packet loss in 105 packets. mMTC service is signified for massive low-cost, battery powered, and low data rate IoT devices. The requirements as specified by the ITU include 10-year of battery life for normal use, low cost devices, less than 10 seconds of latency per 20B packet, and support for up to 1 million IoT devices per square kilometer (Rappaport et al., 2017). Figure 10.3 summarizes the evolution of wireless communication over past years.

## 10.3   APPLICATION OF TELECOMMUNICATIONS IN COVID-19 SCENARIOS

With the advent in technology the power of the network and the versatility of the tasks that can be performed remotely on that network have increased many folds. This paves a path for various applications in education, health, agriculture, and industry. The following section discusses and details the impact of COVID-19 on the specific sectors and provides details of the aid that can be provided by telecommunications for the same.

### 10.3.1  EDUCATION

Various major impacts of COVID-19 on education are closure of schools and colleges, delaying of exams (like JEE, NEET), and postponement and cancellation of exams. Many students from lower classes are pushed towards labour due to the economic impact of COVID-19. 5G technologies will enable very high data download rates diminishing the data buffering delays to non-existent. The connection will be a high-capacity superresponsive link. This will form a foundation of connected devices where everything including mobile phones will be connected. These changes will bring a totally different experience to learning. The introduction of highly responsive AR and VR to education will help remote study enabling mutual interactions and group learning in the online medium. With the development of interactive mediums with AI-based help, available students will have around the clock support availability (Keegan et al., 2005).

**Digital Learning** Digital learning helps students improve on his/her weak subjects or topics. There are numerous open sources for studies. Those students who passed out this year can go for online counseling sessions through PC. According to recent statistics, only two states have above 40% rural household internet service (users), while others have only 10–20%.

**Immersive Lessons with AR and VR** The field conditions and the lab environment can be simulated using VR and AR technologies. Students will be able to actually interact with the virtual environment and can simulate a condition or a study. One of the barriers these technologies will overcome is the mutual interactions. The inclusion of interactive interfaces will enhance the learning perspectives of students.

**5G Helps Tutors Save Time** One of the most time-consuming jobs for the tutor is to prepare the setup for learning and keep it consistent in different sessions. 5G technologies will move the burden to the cloud where the tutoring environments can be set up. This will make it easier for the tutor to login and maintain consistency. Mundane tasks can be automated to save time and resources on 5G. Student-teacher feedback and interaction will become easier.

**Students with Special Needs Can Learn Better with 5G** Learning comes as a bitter challenge to a lot of special needs students. There is a need for some extra assistance for this group of students. 5G technologies will introduce robot apps in the classroom aiding with learning needs. These students will be able to access every learning resource remotely and learn on par with other students (Baratè et al., 2019).

### 10.3.2  HEALTH

Utilization of telecommunications technologies in healthcare has intensified exponentially in the recent past. This has enhanced opportunities for healthcare professionals, patients, and overall management of the healthcare sector. Telehealth provides digital transmission of information that offers advanced working proficiency such as digital medical records systems, continuing education, and research collaboration.

Use of telecommunications techniques has numerous benefits for healthcare providers as well as their patients. People are capable of consulting with their healthcare providers without traveling long distances. Regular monitoring of vital statistics (heart rate, blood pressure, etc.) of people with chronic conditions can be done by sending information to the physician. The results of various laboratory tests are now available online (Russell et al., 2018).

Healthcare providers are benefited in a number of ways such as physical visits of fewer numbers of patients to clinics leading to shorter waiting times, enhanced capacity to provide services to more patients, significant collaborations between healthcare service providers, improved remote monitoring of patients' conditions to provide quality care, etc.

The outbreak of COVID-19 across national and international borders created absolute challenges on traditional healthcare services in India. Transition to telehealth services is certainly appropriate to mitigate health service problems created by the COVID-19 pandemic.

**Telemedicine** Due to the global lockdown because of COVID-19, people have been unable to consult with physicians/doctors physically. This led to the adoption of telemedicine via audio, video, or text to provide remote delivery of healthcare services. Digital health platforms like DocPrime and Practo are ruling India's telemedicine market providing a number of unexploited opportunities. Practo has been observing a surge since the beginning of the pandemic. They have been witnessing a growth in telecommunications, and there is an increase of about 50% in the number of doctors connecting with Practo.

In rural areas with limited access to medical facilities and healthcare amenities, telemedicine functions as a healthcare provider and brings access to doctors and physicians in these areas. Telemedicine is going to decrease the consultation time and enhance services. By 2025, the telemedicine market is envisioned to reach $5.4 billion in India (Anwar et al., 2018).

**Telesurgery** During the COVID-19 pandemic when healthcare workers are under extreme work stress and mobility is restricted, patients with immediate need of surgery are suffering the most. These patients can be benefited greatly by deploying telesurgery devices over remote areas. These devices require minimum exposure and are less invasive than the traditional methods as they are performed by robots. The advent of 5G technologies is going to play a vital role in enabling telesurgery by introducing ultra-reliable, low-latency communication, reducing the latency of the network from 0.27 to 0.01 second. In addition to this, there will be support of augmented reality and machine recommendations from the network (Zhang et al., 2018).

**Data Portability** In the current scenario when everything is connected and every move you make on the internet is logged and analyzed to provide recommendations, there is an opportunity to bring this kind of record coherence in the field of medicine. Patient history is considered one of the most valuable data but often there is a gap between the availability and accessibility of the data. Fifth-generation networks can provide a common platform to store and retrieve data. The data stored on the cloud will be available to the patient and the authorized healthcare professional. This data can be used to detect the rise in health crises for special patient groups. In the post COVID-19 scenario it could help identify high-risk individuals and groups (Zhang et al., 2018).

### 10.3.3 AGRICULTURE

Introduction of massive machine type communication will revolutionize the world of IoT and sensor networks. It will pave a path for expansion and cost reduction for the devices and the network. Agriculture is going to be one of the industries benefiting largely from the advancements made. The availability of multidimensional data from sensors will enable data-driven practices in the agricultural industry. This will provide an edge in terms of controlling the assets being invested in terms of fertilizers, pesticides, and other resources that go into farming practices. This precise management of resources to enhance the productivity and quality of crops is known as "Precision Agriculture."

Precision agriculture is the practice of management of farm resources and crop management on a micro level. This refers to the monitoring and management of the conditions of soil, nutrients, moisture levels, and other indicators to maximize the yield (Valecce et al, 2019.). The following are a few use cases that will be enhanced by the introduction of 5G technologies in the agriculture industry.

**Data Collection on a Massive Scale** The mMTC provided by 5G networks will enable data collection on a large scale, which was never possible before. The inflow of data will drive the decision-making process and will provide an opportunity to automate the recommender systems. These systems can make a forecast for crop yield to drive market demand and supply chains.

**Quality Mapping for Crops** The sensor networks and IoT will be used to monitor the quality of crops. The availability of low power consumption devices will make it possible to install a device in the field to measure crop quality based on various parameters. The health monitoring of crops and the yield maps will give better insight into smart farming.

**Yield Forecast and Monitoring** The sensor networks will be used to train to forecast. This will enable the farmers to preemptively plan their storage and give a chance for procurement in early stages. Planning and better distribution of the crops yielded will make it possible to route the product where it is in demand to increase profits (Razaak et al., 2019).

**Water Resource Management** One of the major issues facing farmers during the growing season is the availability of irrigation sources. The unavailability of up-to-date data and statistics about the reserve waters and the inflow from various irrigation sources can hurt the yield. The information about the current reserves and flows of irrigation water can give an edge to farmers allowing them to plan their irrigation cycle. This will not only increase the yield of the crops but will also lead to lower water consumption due to the fact that the irrigation cycles will be data-driven (Boursianis et al., 2010).

### 10.3.4 INDUSTRY

Production and processing industries are one of the worst hit by the COVID-19 pandemic. There is a shortage of workers and even if workers are available it is very difficult to maintain social distancing norms. Smart applications based on telecommunications networks can help automate the processes and provide remote access to the machinery. A number of sectors can be augmented heavily by the smart networks and telecommunications can play a major role in the same. The following discussion describes a few application areas that the industry can use and profit from.

**Smart Mobility** Smart Mobility includes all the aspects of transport; these can be rudimentary tasks of route planning to the cutting edge technologies involving intelligent transport systems. The outcomes of these applications will be traffic balancing, efficient transport facilities cost reduction, and reduction in emissions. This kind of connectivity among the different verticals require low-latency, high-speed traffic routing. The mainstay for vehicular networks will be the reliability, security, and latency of the network. 5G in its inherent design provides all the facilities under a single umbrella (Marabissi et al., 2019).

**Smart Energy** This involves the management of power plants, the power grid operation and maintenance, power-failure detection, smart charging ports for electric vehicles, and energy savings for homes and industries. The introduction of 5G will enhance the granularity of the network and introduce a huge number of data collection points, thereby increasing the efficiency and availability of the power plants and the distribution systems. The core requirements of this type of application will be reliability, security, and privacy (Leligou et al., 2018).

**Industrial and Manufacturing IoT** The next generation of IoT is about to restructure the landscape of connected devices altogether. It will revolutionize the machine-to-machine communication and intelligent manufacturing and 3D printing. The applications will not only benefit the industry but will make a lasting impact on society. The major requirements of this type of application will be critically high reliability, ultra-low latency, and massive deployment support with security and privacy (Cheng et al., 2018).

**Consumer Applications** On the edge of immersive experiences, 5G will play the role of an enabler with its high bandwidth capacities including consumer applications like availability of 4K videos on phone-augmented reality and virtual-reality experiences. Blockchain-based financial technologies and holographic technologies all also require the service quality of 5G to enable them (Lema et al., 2017).

## 10.4 CONCLUSION

Telecommunications technologies have acted as enablers since the inception of rudimentary systems. It has made its mark in almost every field, and has made significant changes in the way business and personal life is conducted. With the introduction of mobile and wireless networks, the phone became a truly personal and portable thing. Initial wireless technologies were more concerned with voice communications and mobility only. But with the advent of 4G and LTE networks phones have become truly connected devices. With the focus on data communication, the latest technology has made it possible for the devices to connect to other devices on their own. Introduction of IoT and virtual reality has made a long-lasting impact on applications related to health, education, agriculture, and industry. The introduction of 5G technologies will act as an enabler for these underlying applications by providing three mainstays for communication: Ultra-Reliable, Low-Latency Communication (URLLC), Enhanced Mobile Broadband (eMBB), and Massive Machine Type Communication (mMTC).

During the Covid-19 pandemic when social distancing is a norm, 5G technologies will bring remote locations to users by means of high-bandwidth availability, low-latency communication, and massive deployment of IoT devices. AR and VR technologies will play a major role in human interactions. Education infrastructures will benefit a lot from interactive and more immersive experiences. The health sector will see a paradigm shift in personal care and diagnostics. AI and robot-assisted healthcare will be a commonplace activity due to 5G-enabled networks. UAV-assisted monitoring and sensor networks will make it possible for farmers to maintain and produce higher yields by planning their farming activities. Industries will see a change in manufacturing technologies with the introduction of massive machine type communications assisted by AI. 3D printing, smart construction, and intelligent transport systems will be driven by the facilities provided by 5G networks. Thus, the introduction of 5G technologies and applications enabled by it will create a whole environment that will be able to deal with the current situation where minimum contact is needed.

## REFERENCES

Akyildiz, I.F., Gutierrez-Estevez, D.M., Reyes, E.C.: The evolution to 4g cellular systems: LTE-advanced. Physical Communication 3(4), 217–244 (2010)

Anwar, S., Prasad, R.: Framework for future telemedicine planning and infrastructure using 5g technology. Wireless Personal Communications 100(1), 193–208 (2018)

Baratè, A., Haus, G., Ludovico, L.A., Pagani, E., Scarabottolo, N.: 5G Technology and its application oe-learning. In: Proceedings of the 11th Annual International Conference on Education and New Learning Technologies (2019)

Bohlin, A., Gruber, H., Koutroumpis, P.: Diffusion of new technology generations in mobile communications. Information Economics and Policy 22(1), 51–60 (2010)

Boursianis, A.D., Papadopoulou, M.S., Diamantoulakis, P., Liopa-Tsakalidi, A., Barouchas, P., Salahas, G., Karagiannidis, G., Wan, S., Goudos, S.K.: Internet of things (iot) and agricultural unmanned aerial vehicles (uavs) in smart farming: A comprehensive review. Internet of Things, p. 100187 (2020)

Cheng, J., Chen, W., Tao, F., Lin, C.L.: Industrial iot in 5g environment towards smart manufacturing. Journal of Industrial Information Integration 10, 10–19 (2018)

Farooq, M., Ahmed, M.I., Al, U.M.: Future generations of mobile communication networks. Academy of Contemporary Research Journal 2(1), 24–30 (2013)

Gohil, A., Modi, H., Patel, S.K.: 5g technology of mobile communication: A survey. In: 2013 International Conference on Intelligent Systems and Signal Processing (ISSP), pp. 288–292, IEEE (2013)

Holma, H., Toskala, A.: WCDMA for UMTS: Radio access for third generation mobile communications. John Wiley & Sons (2005)

Ibrahim, J.: 4G Features. Bechtel Telecommunications Technical Journal 1(1), 11–14 (2002)

Karapantazis, S., Pavlidou, F.N.: Voip: A comprehensive survey on a promising technology. Computer Networks 53(12), 2050–2090 (2009)

Kaur, G., Kaur, J., Aggarwal, S., Singla, C., Mahajan, N., Kaushal, S., Sangaiah, A.K.: An optimized hardware calibration technique for transmission of real-time applications in voip network. Multimedia Tools and Applications 78(5), 5537–5570 (2019)

Keegan, D.: The incorporation of mobile learning into mainstream education and training. In: World Conference on Mobile Learning, Cape Town, vol. 11 (2005)

Leligou, H.C., Zahariadis, T., Sarakis, L., Tsampasis, E., Voulkidis, A., Velivassaki, T.E.: Smart grid: A demanding use case for 5g technologies. In: 2018 IEEE In-ternational Conference on Pervasive Computing and Communications Workshops (PerCom Workshops), pp. 215–220, IEEE (2018)

Lema, M.A., Laya, A., Mahmoodi, T., Cuevas, M., Sachs, J., Markendahl, J., Dohler, M.: Business case and technology analysis for 5G low latency applications. IEEE Access 5, 5917–5935 (2017)

Maeder, A., Ali, A., Bedekar, A., Cattoni, A.F., Chandramouli, D., Chandrashekar, S., Du, L., Hesse, M., Sartori, C., Turtinen, S.: Ascalable and flexible radio access network architecture for fifth generation mobile networks. IEEE Communications Magazine 54(11), 16–23 (2016)

Marabissi, D., Mucchi, L., Fantacci, R., Spada, M.R., Massimiani, F., Fratini,A., Cau, G., Yunpeng, J., Fedele, L.: A real case of implementation of the future 5G city. Future Internet 11(1), 4 (2019)

Mousa, A.M.: Prospective of fifth generation mobile communications. International Journal of Next-Generation Networks (IJNGN) 4(3), 1–30 (2012)

Patil, C., Karhe, R., Aher, M.: Review on generations in mobile cellular technology. International Journal of Emerging Technology and Advanced Engineering 2(10), 901–912 (2012)

Rappaport, T.S., Xing, Y., MacCartney, G.R., Molisch, A.F., Mellios, E., Zhang, J.: Overview of millimeter wave communications for fifth-generation (5g) wireless networks – with a focus on propagation models. IEEE Transactions on Antennas and Propagation 65(12), 6213–6230 (2017)

Razaak, M., Kerdegari, H., Davies, E., Abozariba, R., Broadbent, M., Mason, K., Argyriou, V., Remagnino, P.: An integrated precision farming application based on 5G, UAV and deep learning technologies. In: International Conferenceon Computer Analysis of Images and Patterns, pp. 109–119, Springer (2019)

Russell, C.L.: 5G wireless telecommunications expansion: Public health and environmental implications. Environmental Research 165, 484–495 (2018)

Samukic, A.: UMTS universal mobile telecommunications system: Development of standards for the third generation. IEEE Transactions on Vehicular Technology 47(4), 1099–1104 (1998)

Tagg, J.P., Mcewan, A.D.: System for providing mobile voip (2015). US Patent 9,049,042

Valecce, G., Strazzella, S., Grieco, L.A.: On the inter play between 5g, mobile edge computing and robotics in smart agriculture scenarios. In: International Conference on Ad-Hoc Networks and Wireless, pp. 549–559, Springer (2019)

Zhang, N., Yang, P., Ren, J., Chen, D., Yu, L., Shen, X.: Synergy of big dataand 5G wireless networks: Opportunities, approaches, and challenges. IEEE Wireless Communications 25(1), 12–18 (2018)

Zhang, Q., Liu, J., Zhao, G.: Towards 5G enabled tactile robotic tele surgery. arXiv preprint arXiv: 1803.03586 (2018)

# 11 Technological Innovations and Challenges in Higher Education during Covid-19

*Rajesh Kumar[1*], Parveen Goyal[1], and Harish Kumar Banga[2]*
[1] UIET, Panjab University, Chandigarh, India
[2] National Institute of Fashion Technology, Mumbai, India
rajeshmadan@gmail.com, pgoyal@pu.ac.in, drhkbanga@gmail.com

## CONTENTS

## 11.1 INTRODUCTION

Since standards like social distancing and self-isolation have become part of the new normal, technology has become even more necessary in the world for better and hassle-free living and education is the backbone of technology. There was good growth in the area of E-learning even before the Covid-19 impact and online learning has boosted the technological advancements to a great extent in this pandemic era. Classes using various online platforms not only offer flexibility but also provide ease and user-friendly adjustable pace.

Moreover, the focus is to enhance the knowledge and skills through interactive audio-visual and Information and Communications Technology (ICT) tools. The instructions are brief and mixed with different applications like tests, surveys, etc. There is a great scope where such platforms can increase the user's participation by voluntary rather than compulsory means (Osguthorpe and Graham, 2003; Anderson, 2010; Ghirardini, 2011).

The evaluation process is also much easier and simpler to develop grades in one way, but may not reveal actual potential/knowledge sometimes. However, the technology opens ways to encourage students to engage equally and to use chat options for asking questions. Governing bodies like UGC/AICTE have created ways to organize various courses/programmes and shared corresponding links to academicians/researchers for knowledge and skill upgradation.

## 11.2   ONLINE LEARNING ENHANCEMENT

Challenges and opportunities during this time have resulted in great innovations for sustainability. Online education has been implemented even as a part of curriculum to upgrade knowledge of students/leaners through interactive tools. A large number of options are available like flipped classrooms, blended learning, massive open online courses, etc. MHRD and UGC provided great initiatives for online education on digital platforms for students/teachers/researchers for enhancement of learning.

A large number of projects has been initiated by MHRD to assist students, scholars, and teachers (www.education.gov.in, www.ugc.ac.in). Details on these online projects covering educational requirements ranging from young to postgraduate students are available on various educational/government sites in the public domain and details about the same have been summarized in Figure 11.1.

## 11.3   TECHNOLOGICAL INNOVATIONS DURING COVID

Not only engineers and doctors, but students also played a great role in making the best use of technologies during this time. A new shift in learning techniques has created a lot of opportunities but at the same time proper internet access, adequate knowledge, and appropriate tools are required for this innovation.

Beyond online learning, various low-cost technological solutions have been offered by students/ scholars especially for face masks and sanitization.

### 11.3.1   Sanitization/Spraying Machines

While moving out for essential work is necessary, several types of sanitization/spraying devices have been developed to protect people from the virus. Since spraying of disinfectants on humans is not recommended (Li et al., 2020; Rai et al., 2020), an organic disinfect tunnel was also developed in this regard.

Engineering students have developed touchless hand sanitizer for one exam centre also. Food-operated and touchless spraying machines have been developed as per the custom needs (Nichols et al., 2007; Srihari et al., 2020; Mathur et al., 2011).

A touchless washbasin allowing the soap and water to be dispensed was also developed by three engineering students. A special sterilizing chamber was also planned for barber shops and salons to ensure virus inactiveness. A two-chamber system, i.e., atomization chamber and tunnel shock chamber, were developed to remove virus hidden in clothes at various places like malls, hospitals, and other places (Maurya et al., 2020).

### 11.3.2   Protecting Mask

A number of start-ups have jumped into the production of masks. Certain factors like ease to wear the mask, breathability, and functional performance have been taken into account for mask fabrication (Novak et al., 2020). Transparent face masks were developed to get full face visibility, which is also needed for people communicating by lip reading (Chua et al., 2020). A multiuse biosafe 3D printed face shield was developed for easy sterilization and good protection (Sapoval et al., 2020; Swennen et al., 2020).

### 11.3.3   Ventilators

To aid medical infrastructure, low-cost and emergency ventilators were developed during the pandemic, which are more user friendly in nature. Development of portable ventilators was also a

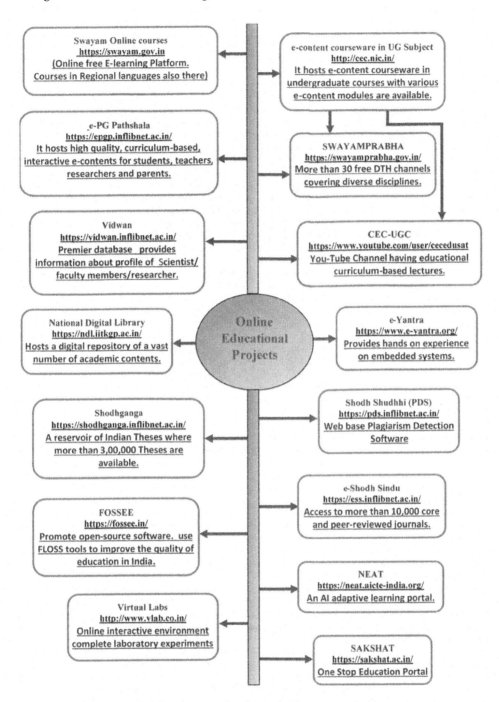

**FIGURE 11.1** Online educational projects ranging from school to post graduates.

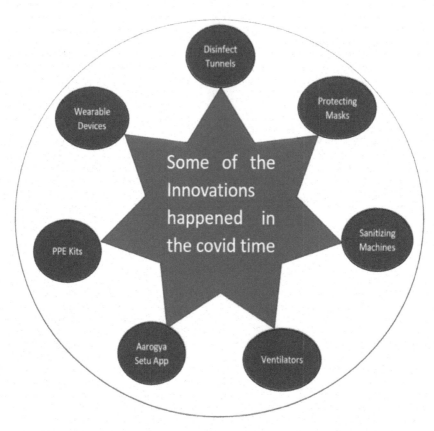

**FIGURE 11.2**   Technological solutions offered during the Covid time.

significant achievement. 3D printers have also been used produce respiratory ventilators (Iyengar et al., 2020; Belhouideg et al., 2020). This pandemic also saw a paradigm shift in patient monitoring to reduce exposure.

### 11.3.4   Arogya Setu App

For contact tracing, the Arogya setu app was developed. This app records the details of people in one's vicinity during day-to-day activity and provides alerts about nearby Covid-infected people (Kodali et al., 2020; Agarwal, 2020; Sriram et al., 2020).

### 11.3.5   Other Developments

Electronics and mechanical engineers designed a healthcare trolley equipped with wash-basin and dustbin for minimum interaction during the delivery of essential items (Peri, 2020). To check the handwashing performance/efficiency, especially for restaurants and hotels, a scanner was developed to check the quality of the hand washer that is later used for handwash performance. For hot and humid conditions, PPE kits were developed made from a special fabric having high breathability. An Arduino-based automated touchless doorbell was developed having ultrasonic sensors to ensure that the person need not touch the bell and virus infection can be avoided. Other than that, sunglasses have been developed to maintain a fixed distance between people in public places. Wearable devices were also developed like a hand band where the hand can be sanitized frequently with just a gentle tap (Mishra et al., 2020; Josephine et al., 2020; Samal et al., 2020). During this challenging time, a few important innovations emerged in the form of technological solutions as summarized in Figure 11.2.

## 11.4  CHALLENGES

Almost every person has felt the impact of this pandemic in one or the other. Access to the internet and technological equipment have been seen as major concerns for online learning. Enhanced academic performance of students needs to be analyzed and correlated with previous performance. While the pandemic at some time resulted in shortage of many necessary items like masks, PPE kits, and ventilators, this also created opportunities to provide innovative solutions to these challenges (Saran et al., 2020; Mathew et al., 2020).

## 11.5  CONCLUSIONS

The pandemic created the need for students and researchers to use innovation and technology to create sustainable solutions. The crisis provided a trigger for thinking differently. The innovations discussed in this chapter can make a big difference to the world in this pandemic. Most of the solutions provided by researchers/students are new, simple, and low cost. The pandemic has offered an opportunity also for more unity and toughness and even more inventions are on the way to improve human life.

## REFERENCES

Agarwal, S. 2020. Aarogya Setu App primed to take lead in COVID-19 war. *The Economic Times* (20 April).

Anderson, J. 2005. IT, e-learning and teacher development. *International Education Journal* 5(5):1–14.

Belhouideg, S. 2020. Impact of 3D printed medical equipment on the management of the Covid 19 pandemic. *The International Journal of Health Planning and Management* 35:1014–1022.

Chua, M. H., Cheng, W., Goh, S. S. et al. 2020. Face masks in the new COVID-19 normal: Materials, testing, and perspectives. *Research* 1–40.

Ghirardini, B. 2011. E-learning methodologies: A guide for designing and developing e-learning courses. *Food and Agriculture Organization of the United Nations*. www.education.gov.in/sites/upload_files/mhrd/files/upload_document/Write_up_online_learning_resources.pdf

Iyengar, K., Bahl, S., Vaishya, R., Vaish, A. 2020. Challenges and solutions in meeting up the urgent requirement of ventilators for COVID-19 patients. *Diabetes & Metabolic Syndrome: Clinical Research & Reviews* 14(4):499–501.

Josephine, M. S., Lakshmanan, L., Resmi, R. N., Visu, P., Ganesan, R., Jothikumar, R. 2020. Monitoring and sensing COVID-19 symptoms as a precaution using electronic wearable devices. *International Journal of Pervasive Computing and Communications* 16(4):341–350.

Kodali, P. B., Hense, S., Kopparty, S., Kalapala, G. R., Haloi, B. 2020. How Indians responded to the Arogya Setu app? *Indian Journal of Public Health* 64(2):228–30.

Li, D., Sangion, L. L. 2020. Evaluating consumer exposure to disinfecting chemicals against coronavirus disease 2019 (COVID-19) and associated health risks. *Environment International* 145:106108.

Mathew, J. L., Mathew T.L. 2020. Invention, innovation, and imitation in India— Necessity arising from the COVID-19 pandemic. *Annals of the National Academy of Medical Sciences* 2:77–86.

Mathur, P. 2011. Hand hygiene: Back to the basics of infection control. *Indian Journal of Medical Research* 134(5):611–620.

Maurya, D., Gohil, M.K., Sonawane, U. et al. 2020. Development of autonomous advanced disinfection tunnel to tackle external surface disinfection of COVID-19 virus in public places. *Transactions of Indian National Academy of Engineering* 5: 281–287.

Mishra, T., Wang, M., Metwally, A. A. et al. 2020. Pre-symptomatic detection of COVID-19 from smartwatch data. *Nature Biomedical Engineering* 4(12):1208–1220.

Nichols, S. K. 2007. Touchless technology helps facilitate infection prevention best practices. *Infection Control Today.*

Novak, J. I., Loy, J. 2020. A quantitative analysis of 3D printed face shields and masks during COVID-19. *Emerald Open Research* 2(42):1–14.

Osguthorpe, R. T., Graham, C. R. 2003. Blended learning systems: Definitions and directions. *Quarterly Review of Distance Education* 4(3):227–234.

Peri, D. 2020. Army EME corps gears up for automation. *The Hindu.* www.thehindu.com/news/national/army-eme-,corps-gears-up-for-automation/article29674740.ece

Rai, N. K., Anushruti, A., Akondi, B. R. 2020. Consequences of chemical impact of disinfectants: Safe preventive measures against COVID-19. *Critical Reviews in Toxicology* 50(6): 513–520

Samal, P., Roy, S., Chauhan, A., Chauhan, V., Malhotra, V., Pundir, S., Singh, S., Kanojia, J. 2020. *Journal on Today's Ideas – Tomorrow's Technologies* 8(1):27–33.

Sapoval, M., Gaultier, A. L. et al. 2020. 3D-printed face protective shield in interventional radiology: Evaluation of an immediate solution in the era of COVID-19 pandemic. *Diagnostic and Interventional Imaging* 101(6): 413–415.

Saran, A., Gurjar, M., Baronia A. K., Lohiya A., Azim, A., Poddar, B., Rao, N. S. 2020. Personal protective equipment during COVID-19 pandemic: A narrative review on technical aspects. *Expert Review of Medical Devices* 17(12):1265–1276.

Srihari, M. M. 2020. Self-activating Sanitizer with Battery Imposed System for Cleansing Hands. Second International Conference on Inventive Research in Computing Applications (ICIRCA), Coimbatore, India, pp. 1102–1105.

Sriram, C., Mohanasundaram, V., Sriram, C., Mohanasundaram, V. 2020. Efficacy of state intervention and public response in the containment of covid-19: A note on aarogya setu. *Journal of Critical Reviews* 7(18):3435–3442.

Swennen, G. R. J., Pottel, L., Haers, P. E. 2020. Custom-made 3D-printed face masks in case of pandemic crisis situations with a lack of commercially available FFP2/3 masks. *International Journal of Oral and Maxillofacial Surgery* 49(5):673–677.

https://www.ugc.ac.in/pdfnews/1573010On-Line-Learning---ICT-initiatives-of-MHRD-and-UGC.pdf

# 12 Shift in Technology
## A Mechanical Engineer's Perspective in the Post COVID-19 World

*Jatinder Madan*

Chandigarh College of Engineering and Technology (Degree Wing)
Chandigarh, India

Jatinder.madan@gmail.com

## CONTENTS

## 12.1 INTRODUCTION

The pandemic caused by COVID-19 has immensely affected the way we live and work. The majority of geographical regions, sectors, society, industry, and technologies have been affected by the pandemic. It is very likely that we will now have to learn to live with the COVID-19 and its various strains; their elimination or eradication does not seem likely to be a reality in the near future. Although vaccination and various other measures have dampened the effect of the virus, its eradication remains distant. Even if the virus is eventually eradicated, the behavioral changes that continue to evolve with each passing day are bound to cause long-lasting effects in the attitudes of our societies and hence shifts in technology usage.

Technological shifts happening due to the pandemic in diverse fields are difficult to comprehend in a chapter. Therefore, this chapter discusses shifts happening in diverse technological fields due to the pandemic from the perspective of a mechanical engineer. Four important aspects in the technology shift influencing the role of mechanical engineers in today's scenario have been discussed in the chapter, namely, digital world, ventilation systems, workplaces, and autonomous systems. The chapter also discusses important technological areas for the mechanical engineers to align with for exploring their multi-disciplinary roles in the industry and the society.

## 12.2   MECHANCIAL ENGINEER'S ROLE IN THE DIGITAL WORLD

The pandemic pushed the general behavior of individuals, industry, society, and even governments towards the *digital*, which appears to be a buzzword now. Therefore, a general opinion emerging out of many forums is that the world is going to be digital. What does this mean? Does it mean that the conventional roles mechanical engineers play will become obsolete and not required by the industry? Will there be a drastic reduction in the requirement of mechanical engineers in the industry? Such questions are regularly pricking in the minds of budding mechanical engineers. The following discussion is an attempt to answer such questions and the anxieties mechanical engineers are now facing.

### 12.2.1   DIGITAL ENGINEERING

The term digital engineering coined not long ago has been there for a while even before the pandemic. Think of the information flow happening for printing of a newspaper. Everything is digital until the soft copy of the newspaper hits the printing machine to bring out its printout for further distribution. In the mechanical engineer's world also digital engineering has been there for quite some time. Digital data now is not only seen as an aggregate of information but has evolved into on-demand conceptualized insight into how systems operate. Engineers can now explore virtual replicas of systems to predict behavior, manage resources, and operate systems more efficiently (ASME, 2020). However, what the pandemic has done is to accelerate the development of new technologies and adoption of existing ones that help design, manufacture, and maintain products with the help of information that depends less on paper and more on digital. Figure 12.1 provides a snapshot of a mechanical engineer working from home benefitting from digital engineering.

### 12.2.2   DIGITAL TWINS

*Digital twin,* which appears to be a derivative of digital engineering, is a new buzzword in the industry in general and mechanical engineering industry in particular. Digital twin is about creating a digital replica of the physical or mechanical process taking place in the industry. The digital replica of a physical product provides much needed information about the process in real time helping simulate other activities and processes for management control and decision making. Perhaps it is also visualized that the data and information that will be generated from the digital twin will help improve the process greatly helping in research and innovation. The pandemic has forced many engineers into work-from-home situations. Digital engineering tools that support distributed collaboration are being considered more thoroughly than ever before (Thilmany, 2020). Figure 12.2 depicts a digital model of a complex engineering product intended to be used for analysis and problem solving.

### 12.2.3   VIRTUAL REALITY

*Virtual reality (VR)* comprising four verticals, namely virtual world, immersion, sensory feedback, and interactivity, is expected to be used effectively for training in diverse processes. VR helps train domain-specific manpower in a virtual environment eliminating (or minimizing) the social contact between the trainer and the trainees. This also ensures that training of technical manpower becomes

**FIGURE 12.1**  A mechanical engineer working from home using digital engineering for designing activity during the pandemic.

**FIGURE 12.2**   Physical model and a digital replica of a turbojet engine for analysis and problem solving.

**FIGURE 12.3**   VR-based welding simulator in use for training, measurement, and analytics.

more efficient, foolproof, and cost effective since unnecessary expenditure on travel, stay, and arranging a physical meeting is avoided. Furthermore, specialized trainers are often assigned with other time-bound activities due to which they need to spare their valuable time for training; VR helps relieve that constraint on training of individuals as well.

It needs to be emphasized that VR allows the use of 3D modeling tools and visualization techniques to simulate physical environment for understanding, training, analysis, and improvements. The developments both in enabling software and hardware have made it possible for the mechanical engineer to take advantage of the technology to simulate physics of a variety of processes. For example, machining of a metal piece on a computerized numerical control (CNC) and welding to join two metal pieces can be simulated in a VR system to help provide realistic feedback about touch and feel to the user. Figure 12.3 provides a snapshot showing VR-based welding simulator in use for training.

In addition to manufacturing processes, other operations where humans are involved can also be simulated and their interactions with the outside environment visualized virtually. For example, assembly of a product can be simulated and interaction of workers, machines, and tools with their neighboring environment can be understood. VR systems therefore provide a number of opportunities for mechanical engineers both in terms of development of enabling technologies and their applications as well.

## 12.3   VENTILATION SYSTEMS

*Ventilation systems* are one of the most important requirements that have taken on new dimensions during the pandemic. A number of cases has been reported in the press where conventional ventilation systems have caused multiple people to be infected. In this context, a ventilation system may be called conventional when air is circulated within the controlled environment without counting airborne virus spread as a design consideration. The ventilation systems even today continue to be designed primarily keeping in view the heating load calculations. However, the pandemic has caused a change in the design thinking to consider virus prevention as a factor. Therefore, lot of opportunities have emerged for mechanical engineers in the field of ventilation system design, simulation, and development of enabling technologies, which help prevent spread of virus.

### 12.3.1 Simulation of Ventilation Systems

Simulation of ventilation systems will help in designing flow of air in such a manner that the air exhaled by an individual is directed for filtration rather than direct circulation in the confined space. NIST[1] developed a free online simulation tool that helps decrease the concentration of aerosols containing the novel coronavirus in the hospital rooms of COVID-19 patients and other spaces such as offices, retail stores, and residences, potentially reducing the likelihood of building occupants becoming infected (Dols et al., 2020). An illustration of the FaTIMA tool developed by NIST illustrating its functioning is shown in Figure 12.4 (Griffin, 2020).

### 12.3.2 Personalized Ventilation System

Ventilation systems of organizations that cater to a large number of employees now cannot depend on conventional ventilation systems. Personalized ventilation systems are expected to be individual centric by supplying conditioned outdoor air close to the breathing zone of each occupant, which has several advantages over conventional ventilation systems (Chen et al., 2012). An illustration of such a ventilation system is provided in Figure 12.5. Furthermore, we need suitable tools that help customize ventilation systems for every building, whether it is a workplace, multiplex, hotel, or a mall.

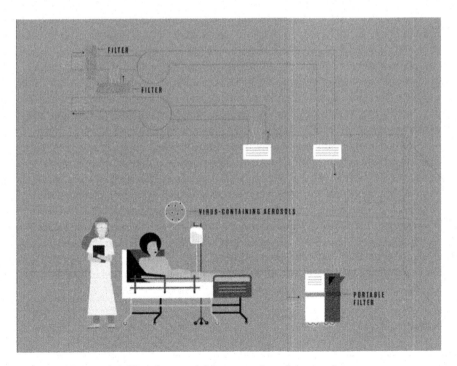

**FIGURE 12.4**  Depiction of NIST airflow model for prevention of corona virus.
*Courtesy*: National Institute of Standards and Technology (NIST).

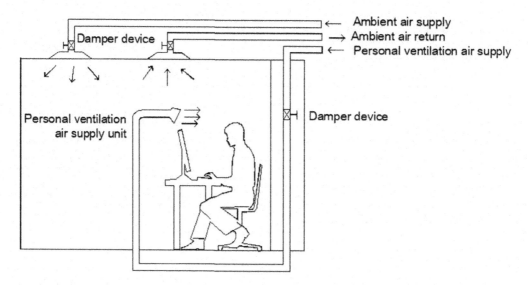

**FIGURE 12.5**   A snapshot of a personalized ventilation system.

### 12.3.3   AIR FILTRATION SYSTEMS

*Air filtration systems,* which are primarily designed to filter the air with an objective to remove the pollutants, need further modification in their design so that aerosols are removed and spread of virus is contained. Therefore, technologies that enable design and manufacturing of such air filtration systems need to be developed. Furthermore, applications that help customize design of ventilation and air filtration systems that cater to different domains such as transport, aircraft, multiplexes, and malls need to be developed. This will require mechanical engineers competent to develop such solutions (or applications) as well as trained manpower to make use of these applications.

Air filtration and ventilation systems for big places such as malls, aircraft, malls, multiplexes, and workplaces will require a good amount of time for their design, development, and installation. Furthermore, such systems will also require lot of capital investment and long gestation periods. Therefore, many people will prefer carrying individual air-filtration system with them, which can keep them from contacting the virus. An individual having such an air filtration system, which is portable, easy to handle, and cost effective, will feel more secure being in public places and large gatherings and travelling. In fact, a few companies have launched such devices (Su-hyun, 2020). Innovation in design, development, and manufacturing of such systems is the need of the hour and abundant opportunities lie in this space for mechanical engineers.

*Ventilation systems for patients*: Demand for low-cost personalized ventilation systems meant for patients facing breathing issues and lack of oxygen has also arisen due to the pandemic. A number of researchers and innovators have designed and developed ventilation systems that are low cost and manageable in the hands of individuals. However, it will require continued efforts from mechanical engineers to evolve innovative designs to cater to patient-specific needs.

## 12.4   WORKPLACES

### 12.4.1   WORKPLACE DESIGN

Future workplaces will need designs that provide social distancing amongst employees as well as employees and customers. Because of the pandemic, a variety of new requirements have emerged that pose fresh challenges for mechanical engineers. For example, opening and closing of doors in many organizations is by application of hands of individuals, which increases the chances of virus spread. Therefore, automated systems will be required that help open and close doors.

Another challenge is the ease of sanitization of buildings, employees, and customers without adversely affecting normal functioning of an organization. Automated systems that provision sanitization of buildings and places in regular intervals or need-based customized intervals will be required. Equipment and furniture that are resistant to sanitizing chemicals are also required so that deterioration of items due to frequent use of sanitizer can be avoided. Furthermore, interfaces, partitions, and furniture that prevent chances of spread of virus will require efforts in developing novel designs and materials.

### 12.4.2 EMPLOYEE ATTENDANCE AND IDENTIFICATION SYSTEMS

Attendance and identification of employees is an important aspect in any organization often used for salary and discipline. Attendance and identification of employees in a physical model had been a common practice in the industry, which more recently has been replaced by biometric systems. However, due to the pandemic, people are reluctant to use conventional as well as biometric attendance system for the fear of contacting or spreading the virus. Figure 12.6 shows a biometric attendance system that has been prevented from use in most organizations throughout the world. Therefore, attendance and identification systems that can fulfill the industry requirement as well as prevent virus spread will be required.

## 12.5 AUTONOMOUS SYSTEMS

### 12.5.1 DELIVERY SYSTEMS

The pandemic has popularized contactless delivery of goods, especially food items. Therefore, organizations are innovating to develop novel delivery systems to face this new challenge. Drones for delivery are a recent concept that has successfully been introduced in many countries including the United States. With such delivery systems, people feel more secure and comfortable while

**FIGURE 12.6** Attendance systems that need finger prints are considered a virus spreader.

**FIGURE 12.7**  Emergence of drone systems for delivery is helping the fight against COVID-19.

receiving their packets. These drones are becoming popular not only for delivery of goods directly at the footsteps of customers but for ready to eat food items too, such as pizzas. Figure 12.7 provides a snapshot of delivery of packets by drones. There is a need to develop custom-designed drones for delivery of goods and food items. For example, quantity and type of payload will require suitable design for environment catering to specific requirements considering parameters such as maintenance of suitable temperature. Mechanical engineers will have ample opportunities in design and development of such systems.

## 12.5.2  AUTONOMOUS ROBOTIC SYSTEMS

Robotics systems are already being used for several applications, which have earlier remained untouched by them. Earlier, robotic systems were used for industrial applications only, mostly in repetitive, monotonous environments or places unsafe for humans. But recently, autonomous robotic systems have been introduced, which can make decisions on their own by taking feedback from emerging situations. These autonomous robotic systems have use in unconventional applications, such as room service in a hotel. Figure 12.8 shows a guest being assisted by such an autonomous robot. The pandemic is going to accelerate use of such autonomous robotic systems for applications, places, and services. For example, the hotel and travel industry will further increase use of robotic systems for better services as well as to maintain hygienic conditions.

## 12.5.3  AUTONOMOUS CARS

Autonomous cars are a recent technological development that has matured to win statutory approval for their use in many countries. Getting statutory approval is critical for this technology as it paves the way for their widespread use passing legal and technological hurdles. The pandemic will further

**FIGURE 12.8**   A robot assisting a guest in a hotel.

**FIGURE 12.9**   A woman using smart phone in autonomous car.

accelerate their development and applications as people prefer driverless cars. Mechanical engineers need to develop enabling technologies and systems by working in inter-disciplinary fields, such as electrical, electronics, computers, and chemicals. Figure 12.9 shows a woman using her smartphone without worrying about driving in an autonomous car.

### 12.5.4   INDUSTRY 4.0, INTERNET OF THINGS, AND CYBER-PHYSICAL SYSTEMS

According to the VDMA (Mechanical Engineering Industry Organization), the majority of the industries are focusing on technology transformations to return to the 'new normal'. The pandemic has triggered the need for a complete overhaul of an organization's technology systems encompassing hardware software, network architecture, and data management affecting people, processes, and technology. Although mechanical engineering industries were already in the process to adopt Industry 4.0, the technology transformation timeline has been positively brought forward with the emergence of the global pandemic. The new normal will therefore require more skilled engineers than ever before. In addition to the core mechanical engineering, engineers need to have interdisciplinary competencies, such as digital engineering, sensors and robotics, industrial automation, data science, artificial intelligence, and machine learning. This aspect of the paradigm shift in expanding the role of mechanical engineering is depicted in Figure 12.10.

## 12.6   CONCLUDING REMARKS

The pandemic has caused irreversible change in society's expectation of mechanical engineers. To remain relevant, mechanical engineers will have to acquire competencies that can be classified into

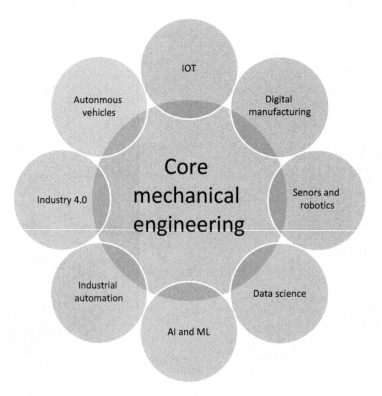

**FIGURE 12.10**   A conceptual view depicting expanding role mechanical engineers in changing circumstances due to the pandemic.

verticals, such as core mechanical engineering fields of thermal engineering, fluids, and strength of materials, and horizontal, such as data science, IoT, cyber-physical system, AI, and ML. Although these changes had been happening for quite some time, the pandemic accelerated the pace of their adoption and development of enabling technologies. Mechanical engineers who have such additional competencies will be more relevant, will have ample job opportunities, and will be better equipped to serve society. However, the comments and discussion do not mean that mechanical engineers with competencies in core mechanical engineering fields will become irrelevant. Of course, they have their own role to play in the development of inter-disciplinary fields, but knowledge of other allied disciplines will provide them an edge to contribute in interdisciplinary teams in a better way. On a positive note, mechanical engineers will continue to play a vital role in variety of technological core mechanical engineering fields and interdisciplinary areas significantly strengthening their role in society.

## NOTE

1 National Institute of Standards and Technology, USA.

## REFERENCES

ASME. (2020). Highlights from ASME's Digital Twin Summit. Retrieved from www.asme.org/topics-resour ces/content/video-the-digital-twin-world-at-your-fingertips (Accessed: April 18, 2023).

Chen, Y., Benny, R., Sekhar, S.C. (2012). Individual control of a personalized ventilation system integrated with an ambient mixing ventilation system. HVAC & R Research, 18(6):1136–1152 DOI: 10.1080/10789669.2012.710059.

Dols, W., Polidoro, B., Poppendieck, D., Emmerich, S. (2020). A Tool to Model the Fate and Transport of Indoor Microbiological Aerosols (FaTIMA) Technical Note (NIST TN), NIST, Gaithersburg, MD, USA [online], https://doi.org/10.6028/NIST.TN.2095 (Accessed April 24, 2023).

Griffin, J. (2020). NIST airflow model could help reduce indoor exposure to aerosols carrying coronavirus. NIST. Retrieved from www.nist.gov/news-events/news/2020/06/nist-airflow-model-could-help-reduce-indoor-exposure-aerosols-carrying (Accessed: April 18, 2023).

Su-hyun, S. (2020). LG Donates Electronic Masks to Severance Hospital. The Korea Herald, July 12. Retrieved from www.koreaherald.com/view.php?ud=20200712000176&np=1&mp=1 (Accessed: April 18, 2023)

Thilmany, J. (2020). Digital engineering during COVID 19. ASME. https://www.asme.org/topics-resources/content/digital-engineering-during-covid-19 (Accessed: April 18, 2023).

# 13 Architecture of Covid Era

*Gurkirpal Singh Bhangoo*

IKG Punjab Technical University, Mohali Campus, Panjab, India

## CONTENTS

## 13.1 INTRODUCTION

Architecture is the art of enclosing spaces for different purposes, uses for habitation, display, exhibit, healthcare, education, security, hygiene, cooking, sale, conferences, offices, sports, banking, and hospitality. It is truly said that "We shape our buildings and in turn our buildings shape us." Spaces we use during our day-to-day functioning have a marked effect on our wellbeing (Alyssa, 2020).

Social distancing is the buzzword in this pandemic, and we need to include this aspect in designing of spaces. The virus stays for 72 hours on any surface, so choice of building materials and surface finishes has an important role in containing spread of disease.

Architecture, as with any other aspect of our daily lives, evolves and depends on socio-economic factors of society. The arrival of COVID-19 will without question also alter the architectural plans of buildings and generate the need for adaption (Nathaniel, 2020).

The COVID-19 pandemic turned the world upside down, and social distancing and halted economies have completely disrupted our lives and will do for months, if not years, to come. But good things have come out of this too. Families have reconnected and embraced slow living principles, parents working from home has reduced travel time and traffic delays, while a renewed focus on health and wellbeing has seen us exercising more, eating better, and connecting with nature. As we now look to life beyond lockdown, the changing values of occupants will impact the way we design homes for a better future.

The pandemic and resulting shutdown will have a lasting impact on how people think about their home, business, and city environments. The first spaces that will be represented in the new design will be the ones that offer more possibilities and until this moment, were still not being completely used and taken advantage of rooftops will be thought of as an area of relation between neighbors, spaces to work out, or even to create urban gardens. Rooftops offer a clear area where people can run, sunbathe, or even have a small barbeque. Rooftops may even eventually be used as drop-off spots for packages by drone. The buildings will look for more natural light and better sunning. Lockdown has made us be more aware of the uses of a balcony, a terrace, or a patio in our homes (Richard, 2020).

DOI: 10.1201/9781003358916-13

'Despite our rapidly changing world, what is evident and consistent is people's desire to feel connected to a larger network. Increased personal space in public spaces is the need of the hour to contain spread of the virus (Hadley, 2020).

Elevators will also become more technological, with features such as facial recognition or contactless buttons. Other changes like this are also expected. In high-traffic environments like offices, retail, and multifamily buildings, touchless options like voice control and keycard swiping will largely eliminate the need to touch buttons and other surfaces. Expect to see greater use of anti-microbial building materials in future designs, like Lapitec, copper, and krion for countertops and bathroom finishes (Regina, 2020).

We will likely see the return of the vestibule mudroom. This will change the open floor trend to include secluded self-disinfecting rooms. Extended periods of isolation will lead to an even more pronounced focus on wellness and longing for connections to nature. Architects will increasingly consider how homes, offices, and other environments can promote wellbeing by incorporating elements of nature.

In the pandemic, architecture has a definite role to play in reorganizing spaces, social distancing, planning spaces, cities, land use, multiple uses of space, work from home space, video conferencing halls, pedestrian cities, online schools, online shopping stores, and IT-enabled warehouses. More open spaces are also needed for people to spread around. Above all special care is to be given to the physical health and mental health of people in the era of lockdown, isolation, social distancing, and quarantine (Peter, 2020).

In this changing and challenging scenario of the COVID pandemic, infection control has been a major concern for control. Our regular building spaces rarely address this issue, so a dedicated approach with special concern for infectious disease control in day-to-day life is needed.

## 13.2 ARCHITECTURE DESIGN FOR COVID

Chandigarh being a well-planned city, taking into consideration each aspect of city for pandemic. Since Chandigarh has been planned using neighborhood units, which is based on localization of resources like markets for day-to-day needs, schools for children within walking distance in sectors. Pedestrian scale has been adopted for city which is the need of the hour for social distancing in era of pandemic. Leisure valley runs across through the city providing healing landscape to the pedestrian masses which is required for physical and mental health of occupants of all walks of life and age groups. The city was planned as a garden city and to be self-sufficient as far as food items are concerned since it has a green belt around its periphery for the sole purpose of providing and fulfilling food, dairy, poultry, etc., needs of the city.

City planning depends on public transport for users again this is requirement of present day to limit public transport. The following components of design are to be taken care of in view of the pandemic to contain the spread of infection:

**Localization.** Strict norms should be put in force to implement concept of localization in below mentioned areas to maintain social distancing, avoid contact, spread of disease.

- Work
- Travel
- Food
- Medical
- Exercise
- Shopping
- Education

### 13.2.1 HOUSE DESIGN

The following concepts should be incorporated into architecture design for the wellbeing of inhabitants:

- Flexibility
- Courtyards
- Medical rooms
- Isolation rooms
- Quarantine
- Gymnasium

**Work from Home**. This concept is very much in tune with preventing the spread of the virus, and design of homes should accommodate the following requirements for smooth functioning of each member of the family and provisioning spaces for these functions:

- Home office
- Home school
- Home theatre
- Home play

**Entrances** to prevent contact in any form are concepts that should be provided for in architecture design of spaces including the following features:

- Scanner-enabled alleys
- Touch-free doors
- Air curtains
- Disinfecting rooms
- Isolation cloak room

### Fixtures

- Hand-washing stands
- Faucets

### Building Materials

- Smooth surfaces
- Non-stick surfaces
- Copper-like materials that control infections

### Electrical Fixtures

- Access activated
- Range sensitive
- Occupant sensors
- Temperature scanners
- Motion sensors

**Home Appliances**. Touchless technology motion sensors or voice commands should be incorporated to operate appliances in our daily functioning to avoid spreading of infection through contact:

- Fridges
- Televisions
- Washing machines

- Fans
- Vents
- Ovens
- Roof mobile chargers
- Fresh air vents
- Air purifiers

**Transport.** The following concepts should be incorporated into transport planning for the wellbeing of city inhabitants:

- Avoiding public transport and long-distance travel
- Cycle paths
- Pedestrian paths

**Food Items.** The following concepts should be incorporated into kitchens for wellbeing of inhabitants:

- Sanitized
- Packed
- Pasteurized

### 13.2.2 HOSPITALITY

**Hotels.** Face recognition check-in, automated security scanning of luggage, e-tagging of luggage, distancing sensors to be put up, along with online booking of rooms, cashless payments online, etc.

- **Rooms.** Automated sterilization, disinfection, indoor air quality sensors to be installed for infection scan, etc.
- **Dining.** Distancing sensors to be put up, indoor air quality sensors to be installed for infection scan, etc.
- **Common areas.** Distancing sensors to be put up, indoor air quality sensors to be installed for infection scan, etc.

**Restaurants.** The following concepts should be incorporated into catering for wellbeing of inhabitants:

- Cashless payments
- Takeaway packets
- Dining areas with social distancing sensors and food dispensing machines
- Indoor air quality sensors to be installed for infection scan

### 13.2.3 EDUCATION

The following concepts should be incorporated into schools, colleges, and universities for safety and hygiene of students and researchers.

**Schools**

- Recording studios
- Lecture relaying stations
- Online classes

**Colleges**

- Recording studios
- Lecture relaying stations
- Online education

## 13.2.4  SPORTS

- Live telecasts
- Isolated gymnasiums
- Telecasting stadiums

## 13.2.5  MASS TRANSPORT HUBS

**Railway stations.** Distancing sensors should be put up in stations, along with online booking of seats, e-ticketing, and cashless payments.

**Airports.** Automated security scanning of luggage, e-tagging of luggage, distancing sensors to be put up, along with online booking of seats, e-ticketing, and cashless payments online.

**Shopping.** Design and layout also has to evolve in response to the pandemic.

**Retail design** merchandise should be shelved away from the reach of buyers in racks only samples to be displayed encased in glass, etc., to avoid touch.

**Showrooms** should be designed in such a way that products to be showcased for window shopping, which will help in reducing contact.

**Parks.** To curb infection spread the following should be provided for users of park premises:

- Water spouts
- Self disinfecting street furniture
- One-way paths

## 13.2.6  AGRICULTURE

**Urban agriculture.** Spaces in a city should be designed for productive landscapes so that landscapes provide required resources like vegetables, fruits, etc., for day-to-day needs of city dwellers.

**Rooftop gardens** can provide effectively for different kitchen ingredients for food preparation and in the process also control temperature inside the building by insulating roofs from very hot or cold climates since plants cover rooftops.

## 13.2.7  TOUCHLESS TECHNOLOGY

Occupancy sensors, motion sensors, or voice commands should be incorporated to operate the following technology in our daily functioning to avoid spreading of infection through touch and along with maintaining social distancing.

- Lifts
- Escalators
- Hotel room entry
- Lighting
- Temperature control
- Ablution
- Wash basins

- Faucets
- Showers
- Cooking ranges
- Washing machines
- HVAC

### 13.2.8  MEDICAL TOUCHLESS

**Hospitals** should incorporate the following for protection of everyone during the pandemic:

- Online appointments
- Entrance controls
- Emergency infection control
- Isolation
- Telemedicine

**E-Governance** touchless

- Bill distribution/payments
- Online notifications
- Online e-passes
- E-certificates
- E-passports
- E-voter cards
- E-affidavits

### 13.2.9  E-BANKING

- E-payments
- Online transactions
- E-deposits
- E-accounts

## 13.3   FUTURE CONCERNS

Our homes will need to change too to make them more energy and heat efficient. While many workspaces, flats, and apartment blocks do not have operable windows, if we are to going to be spending more time indoors, our houses will need to be better ventilated and offer more light. Also we need to avoid something called "sick building syndrome," which is what happens when buildings are entirely sealed and start reticulating pathogens through their systems. Perhaps our homes will even be built to feature "decontamination airlocks."

But as the world grapples with the harsh reality of our current situation, we cannot simply build our way out of the problem, a shift in thinking is needed for any city of tomorrow.

Pandemic architecture is need of the hour for city design in the face of globalized health threats, focusing not just on design, but on our broader relationship with nature. Mankind has abused nature, and generated epidemics, before thinking about new cities, the focus should be on preventing new diseases from emerging in the first place.

So perhaps we should not be picturing shiny new city-center plans when envisioning a pandemic-resilient city. Instead, the changes will be quite practical, like pop-up handwashing stations, and often invisible tracking devices built into our sewers. If we do pandemic preparedness right, our cities might look much as they do today – just a little less crowded, with a little more local open space, and with more of the resources they need to support themselves on the doorstep.

## REFERENCES

Alyssa Giacobbe, "How the COVID-19 Pandemic Will Change the Built Environment," www.architecturaldig est.com/story/covid-19-design as accessed on 15 June 2020.

Hadley Keller, "How the Corona Virus Pandemic Will Shape the Future of Design," www.housebeautiful.com/ lifestyle/a32477800/how-coronavirus-will-shape-the-future-of-design/ as accessed on 15 June 2020.

Nathaniel Bahadursingh, "Ways COVID-19 Will Change Home Design," https://architizer.com/blog/inspirat ion/industry/covid-19-home-design/ as accessed on 15 June 2020.

Peter Maxwell, "How COVID-19 Could Impact Retail Design: Clean Lines, Clean Spaces," www.frameweb. com/news/clean-lines-clean-spaces-covid-19 as accessed on 15 June 2020.

Regina Cole, " Five Ways COVID-19 is Changing the Future of interior design," www.forbes.com/sites/reginac ole/2020/04/17/five-ways-covid-19-is-changing-the-future-of-interior-design/#194ee4a82ee2 as accessed on 15 June 2020.

Richard Moross, "Four architectural designs from history to remember in a post-Covid-19 world Courtyards," https://scroll.in/article/958255/four-architectural-designs-from-history-to-remember-in-a-post-covid-19-world as accessed on 15 June 2020.

# 14 Maps, Digitalization, and Citizenry
## With Special Reference to COVID-19

Vibhash C. Jha

Visva-Bharati University, Shantiniketan, West Bengal, India

vcjha@visva-bharati.ac.in

## CONTENTS

## 14.1 INTRODUCTION

India is a country whose strength lies in its diversity. Diversity in languages, culture, religion, and also geography and climate. It is an unified entity that has perhaps one of the most ancient repositories of ancient scientific knowledge such as ayurveda shastra, astronomy, geometry, and mathematics. This led to development of an advanced civilization on the banks of the River Indus with sound town planning, having a well laid out and organized sewage system alluding to awareness about proper waste disposal. This could only have been possible with profound knowledge about surface slope to facilitate drainage, wind directions to facilitate natural sweeping away of loose dust from the streets, urban housing, and spatial patterns for circulation of enough natural air and light within the individual dwelling units, sites from fortifications to protect from intruders and invaders, etc. Thus, organising of spatial units would only have been possible with all details being planned and mapped before being executed resulting in growth of cities within the civilization.

Even in modern day, the ethos, ancient knowledge, and practices have permeated down to the lives of most Indians in all spheres. Ample examples can be found in building houses, dress, food habits, rituals, classical arts, cultivation of crops, and the rituals followed before sowing, after harvesting, maintaining certain rituals while storing food grains, the profound impact of seasons, and so on. The principles of Vastu Shastra in the building of dwelling units are still adhered too, and refer simply to planning and mapping before physical construction to harness positive energy from nature. Thus, it is noted that the concept of spatial planning, charting of newly discovered land in a rudimentary, manual manner has been prevalent for ages since humankind always wants to be aware of the place he is in or he aspires to move to.

India's diversity, vastness, and richness permeate through its culture, economy, territorial limits, geography, history, and so on. In fact, India's vastness poses a big challenge in governance because as per Census records of 2011 the younger working-age population, ages 18 to 44, represented 112.8 million persons (36.5%). The older working-age population, ages 45 to 64, made up 81.5 million persons (26.4%). Finally, the 65 and over population was 40.3 million persons (13.0%) (Census India, 2022).

For a country that ranks second (i.e., 17.35% of the total world population followed by the United States at 4.39%) the fruits of good governance leading to prosperity should reach every doorstep. Difficult and daunting as it may seem in terms of physical operations of welfare schemes given India's geographical diversity both in physical regions and cultural influences, the quick dissemination of information, aid, and consultancy to all subgovernance nodes from the apex governing centre will rule out and leave no room for mismanagement. The electronic link between the common citizen and administrators should erase any scope for mismanagement, monopoly, and anarchy by a select few. Easy and quick dissemination of knowledge would also mean more awareness among the general populace and also transparent scope for redressal if any mismanagement takes place.

The general aspirations of all sections of citizenry have to be met through extensive and efficient data collection and meaningful utilization of the curetted data for sustainable developmental models in minimum time frames.

## 14.2   RESEARCH DESIGN

India is vulnerable to a wide range of natural hazards, particularly flooding, cyclones, drought, extreme heat waves, landslides, wildfires, and earthquakes brought about by natural causes like climate change/geologic events and also human hazards brought about by separatist violence. It is very necessary to emphasize here that hazards to human life can only be tackled if citizens can be protected from natural disasters. The terrain and climate change factors cannot be conquered but can be tackled with well-planned policies that will lead to development and security for all citizens. It is only then that all citizens will be ethically inspired to shun violence. A successful nation's development lies in the welfare of both the mental and physical well-being of its citizens.

Research and analysis show that, as a result of climate change, the intensity, duration, and frequency of weather-related events such as landslides, deforestation, loss of agricultural lands, reduced crop yields, reduced pasturelands, and forest land reduction due to increase in settlements to house a growing population, floods, etc., are likely to increase. Experts are of the view that droughts can not be treated as natural hazards as they generally evolve from unscientific long-term agricultural practices of man but can be suitably mitigated with proper water-saving techniques of farming and extraction of ground water and storage of rainwater. In India, vulnerability to natural hazards is intensified by unplanned settlements due to high population density and growth in urban and coastal areas.

## 14.3   GEOSPATIAL SYSTEMS AND MAPS

As such, more than ever, "New India" requires prudent use of geospatial technologies for resource mapping, distribution, dissemination, and utilization. For all these parameters to work effortlessly, seamlessly, and in perfect collaboration the spatial (distance, region, area) and temporal (time) frame have to work simultaneously (i.e., fast dissemination of information and communication through efficient use of technology). Maps as we all know today are essential, cost-effective, and thereby indispensable in almost all spheres of our modern life. They give out three-fold information: location, position, and situation (LPS). Location relates to the geographical place, position refers to its status as compared to the world, and situation deals with real-time, ground verifiable truth. Thus, maps are a way to communicate with locational information.

The rapid popularity and demand for maps has increased manifold primarily due to the advent of geospatial technology (remote sensing, geographical information system, and global positioning system) along with digitization of cartographic methods and the growth of internet traffic. The advent of Internet of Things (IoT) systems that collect data, collate, and transfer data and finally allow to analyse data and disseminate on human interface or backend systems or business analytics allowing to take action saves both valuable time and loss of data, which otherwise would have happened with human interference during transfer. The updating of data is also seamless and done in minimal time thereby being cost effective.

Since the early days in cartography, people realized that to be successful they would need information and data from sources beyond their immediate workgroups. People recognized the need for data sharing. Modern cartography emphasizes participation, sharing, and collaboration. It is no longer confined to experts in this domain, but rather it encourages involvement of common people. Geospatially oriented social media communications have emerged as a valuable information resource to support crisis management. Cartography is no longer confined to paper maps. It has shifted to the cloud (large clusters of computer servers with a variety of software installed on them).

"The diagram above explains the geospatial ecosystem in a digital environment. Pillar I of the geospatial ecosystem lists the technology segments that are involved in spatial data collection and analytics, namely GNSS and positioning, GIS and spatial analytics, earth observation, and 3D scanning. The second column lists various platforms that facilitate the outreach of geospatial technologies in their process of reaching out to users, standards, open data, and interconnected systems. The third column stands for the processes on which geospatial rides – business intelligence, digital engineering, and workflow automation. Finally, it is evident to the mediums that are used to deliver the end-to-end solutions – enterprise technologies, social media, mobile apps, and web portals. The common drivers in this entire process are IoT, AI, cloud, wireless and broadband and big data. The end users of the solutions created could be citizens, governments, or private enterprises, who all can be grouped together as geo users" (Datta) (Figure 14.1)

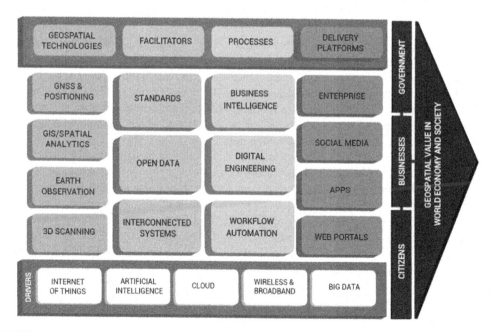

**FIGURE 14.1**   Digitalization-Geospatial Value in World Economy and Society.

Source: www.geospatialworld.net/blogs/geospatial-technologies-in-digital-platforms/

A fast and affordable computing network enables all of us to have access to rich collections of information. Every layer has a URL that makes it searchable and easily usable. Data from different sources can be combined into a web map and can be published as an app that can be shared with anyone. This new era of cartography has a much wider audience and influence on common people.

The latest ability to fly specific missions with UAVs as part of geospatial technology has opened up whole new ways to implement cartography for a new range of problems pertaining to:

1. Assessing traffic conditions
2. Law and order problems
3. Monitoring wildlife habitat
4. Responding to emergencies
5. Responding to natural disasters
6. Monitoring agricultural fields
7. Land degradation and desertification
8. Assessing coastal regions for turbidity and discharge from rivers
9. Tracking aquamarine movement during spawning season
10. Securing national borders

Geospatial technology has totally engulfed every aspect of our lives and this has brought down distances, improved communication speed greatly, increased accountability and alertness, and is helping individuals to store massive quantities of data virtually.

Jack Dangermond of the Environmental Systems Research Institute (ESRI) conveyed the potential of GIS as follows:

"Knowing where things are, and why, is essential to rational decision making.The application of GIS is limited only by the imagination of those who use it." In fact, this statement applies to any technology but more so to any computational technology. A variety of "unintended usages" emerge after adoption of a technology.

## 14.4   CARTOGRAPHY AND DIGITALIZATION

Our country has progressed from the initiation of the first mapping of landholdings for revenue collection since the time of Sher Shah Suri by Raja Todar Mal who rose to fame later when he joined Emperor Akbar's court as his revenue minister in 1560 A.D. Further, even though now in India the land marking system has adopted an electronic format based upon GIS the system and concept of patwaris still remain. Thus, Raja Todar Mal overhauled the whole concept of ownership of landholdings and introduced standard weights and measurements, which was followed throughout Akbar's kingdom and is referred to and consulted even today.

Now to the present times, geospatial technology is no longer a domain out of the reach of the common man. Now the requirement is in diverse fields, cartography requires more precision and ease of use. Learning cartography has always been associated with geography, but that pathway is now diversified. Everyone wants to know where they are, what resources are available within reach, but Indian universities have never realized the need for training students for professional cartography. Yet there is now such a wide demand within our country as well as abroad. Maps so far have been meticulously and authentically produced by the Survey of India and the National Atlas and Thematic Mapping Organisation once on paper mode and now in digital mode. Both of these organisations have contributed at a steady pace to the mapping of Indian states, presenting cartographically different facets of national and regional or socio-economic realities and characteristics.

Cartography has never gained its due ground because of the following:

1. Foremost, it has been treated as a mere tool to illustrate demarcation of space on paper and pointing of locations of some feature based on a particular theme.

2. Secondly, educational institutions have always associated their recognition with geography and never as an independent discipline, which has usage in other fields also.
3. There is also a dearth of institutions in India that are devoted to this discipline.

Cartography, if it is to be treated as a technical discipline that helps us to describe and illustrate the concept of area or regions, can also be taught as subject in our technical institutes, which will create a pool of sound cartographic professionals who can provide service to the required fields with more precision.

At present, professionals are sent to train at our various institutes like IIRS, NRSC, NATMO, or SOI for a certain duration where a general training for a certain period is given and a certificate at the completion of the course. This offers limited scope for adaptation of the knowledge for multiple uses.

In other words, a good cartographer is both a scientist and an artist. A cartographer must have a thorough knowledge of his subject, must be intelligent enough to make the right selection of the features to represent, should make artistic judgment to effectively use lines and colours, and is also expected to be aware of the tricks of effective visual communication. In fact, much like archaeology, architecture, and medicine, art and science are inseparable in cartography. In modern times research on earth studies has become more focused on the impact of natural processes upon those regions that are anthropogenetically important. The need for awareness and the expertise required to tackle problems like drought, floods, soil loss, land degradation and desertification, and gullying in tropical lands are gradually gaining priority among researchers and policy planners today.

Nations have set up task forces to delineate and map, discuss among its experts, and use latest techniques like geospatial techniques (remote sensing, GIS, and GPS) to monitor the problem areas and finally arrive at suitable methods to address these problems. So the requirement in terms of cartographic professionals in the coming times is going to be intense and a country like India can surely not afford to let go of such an opportunity of developing such indigenous talent and also bringing about a paradigm shift in the employment sector.

Cartography has evolved since man started to settle down along the fertile riverbanks and settlements started to sprawl over the adjoining flood-plains. Kingdoms were established and the need arose to expand them. Natural features like rivers, mountains, hills, or deserts were considered boundaries and taken as expanse of a particular kingdom. Then there arose the need to maintain records, collect revenues, and gift or receive land as tokens of appreciation. "Mahajanapadas (Sanskrit: Mahājanapadas), literally "great realms" (from maha, "great," and janapada "foothold of a tribe," "country") were ancient Indian kingdoms or countries. Ancient Buddhist texts like Anguttara Nikaya make frequent reference to 16 great kingdoms and republics (Solas Mahajanapadas) that had evolved and flourished in a belt stretching from Gandhara in the northwest to Anga in the eastern part of the Indian subcontinent and included parts of the trans-Vindhyan region, prior to the rise of Buddhism in India" (Kmusser, 2012).

Thus the requirements for cartography in ancient and medieval India were of three types:

1. Conquest
2. Rule
3. Administration comprising
   a. revenue collection,
   b. describing and estimation of cultivable and fallow land, and
   c. estimation of boundaries for moving military resources and trade.

Maps served as a function of measure, as a means of enquiry, and as a method of examination. It is to be noted that all these exercises did not arise at the same time and gradually evolved over time and as requirements developed. Trade routes already existed over land and through sea routes. The demise of Babur in 1530 A.D. led to the establishment of Sher Shah Suri's empire in 1538 A.D. His system of administration and land reforms provided the foundation for successful management of the Mughal Sultanate later. Raja Todar Mal had already gained experience in land reforms and revenue collection under Sher Shah Suribefrore being inducted into Akbar's court. It can be emphasized that Raja Todar Mal's system of land surveys was his foresight. Lead measuring rope, called *Tenab*, was standardised by joining pieces of bamboo with iron rings so that the length of Tenab did not vary with seasonal changes.

The maps of the Mughal Era and pre-Mughal Era were based on the territorial delineations for e.g. the location of cities, rivers and other physical entities that were derived from the astronomical observations and calculations. These maps were not precise until the East India Company deployed the cartographic technologies used in Europe into their modern form and including the technique of triangulation (also known as "trigonometrically survey") at the beginning of the 19th century, where GTS played a key role as an instrument of the British cartographic control over India. During the reign of the British East India Company from James Rennell's Survey of Bengal (1765–1771) to George Everest's retirement in 1843 as Surveyor General of India, geography served in the front lines of its territorial and intellectual conquest of South Asia (Geospatial World, 2010).

The 18th century under the British rule in India took cartography to another level. The necessity for organized cartography arose in colonial India chiefly due to the expansion of British kingdom and establishment of dominion. Three major types of surveys were carried out in British India:

(i) Topographical
(ii) Trignometrical
(iii) Revenue collection

Surveying and mapping were important parts for administration of newly acquired unknown territories and India was surveyed and mapped at village as well as country level. The first detailed map of India was by a French geographer D'Anville titled "Memoirs" published in French in 1752, which was based on travel routes and rough chart of coasts. The map of Hindoostan by James Rennell in 1788 was based on survey methods and distances to be measured were chained and observations were taken for latitudes and longitudes at certain stations.

The laying of railway lines in the 19th century was yet another reason as it facilitated topographical surveys and recordkeeping in the form of topographical maps. Thus, before mapping was necessary for revenue and military purposes (i.e., to delineate the extent of kingdoms), but then was required for transport and communication purposes also. Railway lines gradually connected the whole of India together with the main cities being recognized as important railway stations.

Maps are the only entity that has use in every field that has to do with movement, real time management, location-based activities, or reaching out to the common citizen and cartographers are the need of the nation now to progress from maps to apps (applications). The advent of remote sensing techniques has enabled us to see beyond what our human eyes perceive, providing new perspectives of the Earth. Satellites have sensors that can measure non-visible information, such as infrared energy, across the electromagnetic energy spectrum that enables us to map and analyze beyond the visible. Thus, modern cartography coupled with big data (i.e., high-volume and high-variety information that requires cost-effective, innovative forms of information processing for enhanced insight and decision-making) can be only successful if young citizens are motivated, trained, and job opportunities created for a pool of cartographers by the government.

## 14.5   GIS AND ITS FUTURE

The role of the National GIS (Geographical Information Systems) is instrumental in the fields of weather, land use, soils, drainage (rivers), admin maps, village sets, roadways, railways, forest regions, groundwater, coastal maps, slope estimation for landslides, or construction in mountainous regions, agriculture, urban areas, disaster management, environmental planning, which need to be constantly updated with real-time data and should be geotagged. GIS is now a tool for total solution of society's problems and prospects.

Some important areas where geospatial technology has applied big data for enhanced analysis include:

- Climate modeling and analysis
- Location analytics
- Retail and E-commerce
- Intelligence gathering
- Terrorist financing
- Aviation industry
- Disease surveillance
- Disaster response
- Political campaigns and elections
- Banking
- Insurance

## 14.6   NATIONAL SPATIAL DATA INFRASTRUCTURE (NSDI)

1. National infrastructure for the availability of and access to organized spatial data
2. Use of the infrastructure at community, local, state, regional, and national levels for sustainable development

It is felt that cartography in our country is a resource lying latent and can act as a master key to opening up the locks of our other resources as it will provide information regarding location, spatial organization, accessibility, and finally the extent of exploitation that can be undertaken. Our youth have a great resource in their hands and it is certain that with a planned road map cartography can be given its due recognition and in a way help to remove the disparity between educated and unemployed by opening up newer avenues of employment.

## 14.7   MAPS TO APPS

If more focus is given on the applications and applicable areas of cartography i.e., from MAPS to APPS like:

1. Web Services and Android applications
2. Orienting GMS to LBS application (Location-Based Services) and also establish a Centre of Cartographic Teaching and Research for:
   - MAPS
   - Digital Data
   - Digitalization
   - Society welfare and establishment of NATIONAL DATA BANK in coordination with the Ministry of Science and Technology
   - Digital Education System in India (DESIN)

India as a nation can suitably bridge the gap between education and employment. The multiple applications of cartography is itself a major deciding factor in its popularity and training of cartographers should be done keeping in mind that they will generate our next generation of dedicated cartographers. It is possible we will see a rise in the number of cartographers who will now map planets of the solar system in addition to our planet. Space probes are presently bringing back planet surface information. Mapping of surfaces of Jupiter, Mars, and Venus have been in progress. Planetary cartography is in a nascent state in our country but cartography will be instrumental in paving the path ahead. Cartography has an assimilating role together with geospatial technology (remote sensing, GIS, and GPS) in building a stronger nation and the immediate imminent need of the hour is to develop an institute that will endeavor in creating professionals trained to use cartography for creating cartographic products that will cover all fields crucial to nation building.

## 14.8   THE POST COVID-19 RESPONSE

By now it is evident that the policy of no physical contact is here to stay. Virtual traffic has increased with homes turning into working space without the pressure of commuting in crowded public vehicles or being stuck in endless hours of traffic. Not only has this been a boon to the environment with noxious fume emitting vehicles being off the road but also good for families getting time to spend more time together. What has come to be of vital importance as a result is the spurt in locational analytics such as risk of catching the virus at a particular location, lockdown areas, and safety measures issued by the governments keeping locational statistics in mind. It can be said that almost all sectors have been affected from medical services, trade, services, education, entertainment, and travel. For educators in a country like ours new distance learning resources and apps have to be made available especially since most education is disseminated either in Hindi or the language of a particular state. It is not that all children attend schools where English is the medium of exchange. Only a few do. Educators have to be suitably trained to help them navigate the transition to distance learning and should be accessible to each and every learner in our country. If children are to be encouraged for distance learning through online platforms to continue social distancing, regional diversity has to be maintained and individual state governments can be suggested to develop their own models of distance learning platforms keeping in mind cultural sensibilities.

However, the reality is that we're really only in Phase 1 of our journey to overcome this pandemic. If you read the American Enterprise Institute's (AEI) report entitled "National corona virus response: A road map to reopening" they talk about four phases:

- Phase I: Slow the Spread
- Phase II: Reopen, State by State
- Phase III: Establish Protection Then Lift All Restrictions
- Phase IV: Rebuild Readiness for the Next Pandemic

To move from one phase to another, we need a set of actions and achievements to be seen as societal indicators – and so many of them can be triggered by the effective use of geospatial technologies. Geography might not have been the most exciting profession in the past few years, but in a space of 3 months it has become one of the most important – especially for those professionals who have a mix of geography and data science expertise – otherwise known as "spatial data science" (Broderick, 2020).

The government has developed the Aarogya Setu App, which takes real-time locational and medical data to alert users about possible COVID-19 carriers from a range of 500 m to about 10 km around. "Using site selection techniques for testing centers is critical in order to serve the highest number of potentially affected citizens – which is where self-assessment app data can come in handy alongside their existing socioeconomic and demographic datasets on the population" (Broderick, 2020). For a country like India it is imperative that the government massively scale contact tracing, isolation, and quarantine. A gradual and sustained reduction in cases for at least 14 days and states able to conduct active monitoring of confirmed cases and those who have come in contact can be suggested. A reduction in cases can simply be seen in the numbers (as long as the data collection and infrastructure is intact and there is no loss of data), but regularly tracking and monitoring confirmed cases and their movement can also be tracked with geospatial technology. It might not always be possible to completely track due to many social issues, but human mobility data along with other location data streams could provide some relevant indicators. Data science teams can look at human mobility data to observe the relative changes in the number of visits to say a grocery store at a local level – which, if we gradually transition out of lockdowns, could be relevant to analyze. This can be suitably used by the government to promote spatial analysis of such medical data based upon cartographic methods.

## 14.9   CORONA-GRAPH, ITS CHARACTERISTICS, AND IMPLICATIONS

Point of Inflexions play a vital role in giving shape to the pandemic, corona-graph (COVID-19). Approach Segment, Rising Segment and Recession Segment, Point of Rise, Formation of Peak and Crest, shape of Crest (conical, rounded, and S shaped) are affected by improper behaviors, policies, population, food, economy, education, media, myths etc. It is noted that various organizations/institutions are concerned about the myths that are essential for restoring the approach segment of the corona-graph. Its characteristics are as single-peaked and multi-peaked. Myths are: (i) human beings in danger will panic and breakdown in hysterics or flee in panic for safety and security; (ii) The helplessness of COVID-19 victims is normal; people are dazed and numb after a sudden pandemic; (iii) This will cause long-term psychological problems and people will be in personal trauma and severe emotional scars; (iv) Anti-social behavior, such as looting, robberies, corruptions, and women crimes are common in pandemics; (v) Home sickness and sustainability.[10] Reduction of population, economic decline, and disasters are the main concerns of COVID-19. A time slot will be needed to recover loss at the local, focal, regional, and global levels.

## 14.10   FUNCTIONING OF THE HEALTHCARE SYSTEM

Healthcare systems are seeing a spurt in the level of demand, with a country like India having to pull together makeshift hospitals or even using railway coaches as hospitals or quarantine ports. Regular inspection of medical facilities including hospitals and primary care clinics will be fundamental to estimate care capacity and vulnerability. To monitor this effectively, health inspectors will need to be equipped with the right mobile apps to do so – with existing staff already overstretched and unable to gather and share data physically in the way they normally would have done under pre-COVID circumstances.

For example, Radboud University in the Netherlands "has mapped vulnerable population groups in Brazilian cities, visualizing a model to estimate future demand for hospital beds using the Huff Model and hospitalization rates. This spatial interaction model calculates gravity-based probabilities of consumers at each origin location. From these probabilities, potential visits can be calculated for each origin location based on disposable income, population, or other variables" (Broderick, 2020).

## 14.11   COMPREHENSIVE COVID-19 SURVEILLANCE SYSTEMS

As mentioned above, maps of populations exhibiting symptoms and the resulting epidemiology analysis is crucially important in order to track the spatial spread of the virus – using both medically tested data and self-assessment data captured by web and mobile apps.

Showing the number of positive cases by pin code (data made available by local health units or hospitals) including additional layers of data such as available local care givers, nearest isolation centers, and medical units can be geospatially interspersed on an interactive interface for all people in the locality thereby creating more awareness. Using GPS to track locations to ensure against quarantine breach, and sending alerts if people leave designated areas is also possible. Using apps in the detection of Covid-19 is useful to the general public and also to help people report their symptoms and to learn about the virus and health responses. Apps can be used to trace contacts through inter-action and proximity analysis. They can also be used as quarantining enforcement tools, monitoring locations and interactions.

Thus, there is a broad scope for using geospatial technology to phase out lockdowns and at the same time in creating awareness, consensus, and alertness among citizens for welfare in general. To combat any such pandemic situation in the future, which will certainly impede economic growth and destabilize our vision for social welfare, development, and equality in health and wealth sharing of key information, bulletins and spatial advent of any pandemic situation or disaster is necessary. This helps to create local response and combat teams in advance with expert training, who can be put to use efficiently in times of crisis.

In the words of Dr. A.P.J. Kalam, "We need to explore how geospatial technology can help the bottom of the pyramid. Creating a sustainable development model for 3 billion rural people involves linking data information exchange and proper dissemination." Thus, the gap between maps to apps can be successfully bridged by motivating and training young citizens. It is they who will carry forward the dream of a "Digital India" and "New India" from local to the global level, leading to products from **Maps to Apps.**

## REFERENCES

ArcGIS. Available from: www.arc gis.com/apps/Cascade/index.html

Broderick, F. (2020). Location Analytics: A Roadmap to Post COVID-19. Available from: https://carto.com/blog/location-analytics-a-roadmap-to-post-covid19

Census India (2022). Office of the Registrar General & Census Commissioner, India; Ministry of Home Affairs, Government of India. Available from: www.censusindia.gov.in

CIA. Available from: Freedom of Information Act Electronic Reading Room. Available from: www.cia.gov/library/readingroom/docs/CIA-RDP08C01297R000200140008-9.pdf

Datta, A. How are digital technologies influencing geospatial technology trends? Available from: www.geospatialworld.net/blogs/geospatial-technologies-in-digital-platforms/

Disasters and social response. *ITC Journal*, 1989, 3/4. Geospatial World (2010). Available from: Mapping An Empire – The Geographical Constructuion of British India (1765-1843). Available from: www.geospatialworld.net/article/mapping-an-empire-the-geographical-constructuion-of-british-india-1765-1843

Jagran Josh. Available from: www.jagranjosh.com/general-knowledge/the-sur-empire Kmusser (2012). Map of India, 600 BCE. Available from: www.ancient.eu/image/321/map-of-india-600-bce/

Stephen A. (2012). Natural disasters in India with special reference to Tamil Nadu. Available from: ww.researchgate.net/publication/265914690_Natural_disasters_in_India_with_special_reference_toTamil_Nadu

# 15 COVID-19 Pandemic and Environmental Sustainability

*Aditya Vikram Agarwal,[1][^] Pawan Kumar,[2] and Rana Pratap Singh[2*]*

[1]HS Srivastava Foundation for Science and Society 04, Eldeco Express Plazas Shaheed Path, India
[2]Department of Environmental Science, Babasaheb Bhimrao Ambedkar University, Raebareli Road, Lucknow, India
*Corresponding author: rpsingh@bbau.ac.in; dr.ranapratap59@gmail.com
[^] adi_10a@rediffmail.com

## CONTENTS

## 15.1  INTRODUCTION

The Coronavirus Disease 2019 (COVID-19) outbreak has emerged as a global emergency. In December 2019, a high number of unusual pneumonia cases of unknown etiology were detected in People's Republic of China. These cases were linked with a local animal market selling fish, poultry, and other animals (Ali et al., 2020). A molecular analysis revealed the presence of a novel coronavirus in samples collected from COVID-19 affected patients, later named severe acute respiratory syndrome coronavirus 2 (SARS-CoV-2). This virus is thought to be transmitted from another mammal to humans and has taken the form of a rapidly spreading global pandemic (Ather et al., 2020).

DOI: 10.1201/9781003358916-15

Chain reaction (RT-PCR) tests have been performed on large scale, giving insights into different properties of this virus. Coronaviruses are a family of single-stranded RNA viruses that have been found to infect a vast variety of animals including humans (Kooraki et al., 2020). In Latin "corona" means crown. There are four different subfamilies of coronaviruses (alpha, beta, gamma, and delta) identified so far. The alpha and beta subfamilies have originated from mammalians, largely bats, while the gamma and delta subfamilies originated from birds and pigs (Ather et al., 2020). The members of the alpha subfamily cause mildly symptomatic or even asymptomatic infections while beta subfamily members are known to cause severe disease and fatalities and SARS-CoV-2 is among the beta coronavirus subfamily. The genome of the virus houses four structural genes that encode spike protein (S), nucleocapsid protein (N), membrane protein (SM), and membrane glycoprotein (M) along with membrane glycoprotein (HE) occurring in few beta-coronaviruses. This genome was found to be 96% identical to a bat coronavirus (Kucharski et al., 2020; Huang et al., 2020), which further strengthened the anticipation that this virus had made its transition from animals in the animal market of Wuhan, China.

SARS-CoV2 enters the human respiratory tract via the mouth, nose, and eyes and after an incubation period of about 0–14 days, primary clinical symptoms of the disease, very much similar to pneumonia and common cold (headache, fever, inflammation in airway, nasal obstruction, sneezing, runny nose, cough, and asthenia), start to appear. Severe symptoms usually include serious respiratory syndrome, kidney failure, and even death. The main target of this virus is lungs as the spikes of virus (binding domains) get attached to the cell receptors of the lungs (Jaimes et al., 2020).

The disease is transmitted from infected to non-infected by nasal secretions and mucosa that can often be controlled, at least partially, by following hygienic measures. During the initial phase of covid, the affected patient received symptomatic therapy, while coronavirus vaccine development research was underway. Few reports from that time period provide descriptions of asymptomatic infections and gastrointestinal symptoms in young children (Li et al., 2020; Zhou et al., 2020).

To date, nearly ten million positive cases have been reported and approximately half a million people have lost their lives because of this deadly viral disease in more than 200 countries across the globe. Datasets currently made available by international organisations like the World Health Organization (WHO) have revealed that the majority of global cases are occurring within 8–10 countries, which are United States, Brazil, Russia, India, United Kingdom, Peru, Chile, Spain, and Italy (https://covid19.who.int/).

The COVID-19 pandemic is impacting all aspects of human societies; however, these impacts are different for individual societies depending on their socio-economic status, cultures, and traditions. There are several reasons different socioeconomic groups are varyingly affected by this pandemic. These socioeconomic reasons include urban and rural settings, population density, education levels and related lifestyle, the size of households, homeowners, tenants, etc. (Messner, 2020). It can be clearly observed from the current global scenario as well as reports from Mumbai-slum (Dharavi) that lower socioeconomic status groups are at higher danger from the spread of this disease. Factors like population density, household size, and social distancing levels directly affect the chances of catching virus. Governments are trying to "flatten the curve," by enforcing border shutdowns, travel restrictions, and quarantine thereby creating a fear of impending economic crisis and recession (Venter et al., 2020).

The pandemic is considered to be aggravated by the mutations caused by environmental degradation and an increasing habitat mixing of wild animals with human settlements due to loss of their natural habitats, domestication, and eating. Environmental factors, pollutants, and radiations are known to cause mutations in lower life forms. The pandemic and its associated slowdown and lockdown of human movements and clustering movement restrictions and clustering of offices, industries, and trading centres have established the need of a new paradigm shift in economic pathways. The work from home format, e-learning, online meetings and communication, and realization of need for a local economy have opened up new challenges and opportunities. We have attempted to analyse and discuss these aspects in this chapter.

## 15.2 THE STEADY INCREASE IN INCIDENCES OF GLOBAL PANDEMIC

In recent decades, increase in frequency of infectious animal-origin diseases also called "zoonotic diseases" have been seen such as SARS, Middle-East respiratory syndrome (MERS), Ebola virus disease, avian and pandemic influenza, and the recently emerged COVID-19 (Ali et al., 2020). This increase in frequency and impact of diseases currently account for approximately 75% of all new human diseases, and it has been anticipated that this number of emerging animals-origin diseases will keep increasing. The basis behind such anticipated risks of zoonotic diseases are anthropogenic activities like loss of forest cover, biodiversity loss, and climate change (Aguirre, 2017).

Globally, many species lost their natural habitat due to various anthropogenic activities such as agricultural intensification and urbanisation transform (Olivero et al., 2017). This results in shrinkage of contiguous natural habitats into smaller, isolated remnant patches surrounded by matrix of human-modified land. Land transformation has cascading ecological effects that influence species persistence, resource availability, community composition of plants animals and microbes, and population carrying capacities, which ultimately influence the transmission of infection diseases between species (Rulli et al., 2017). These "zoonotic spillovers" across core-matrix boundaries has led to outbreaks, declines in populations, pandemics, and even species extirpation (Figure 15.1). Zoonotic diseases represent a global public-health burden and their occurrence, emergence, and re-emergence have shaped the lives of human beings for centuries (Rodriguez-Morales et al., 2020).

Zoonotic spillovers require a conglomerate of various factors to ignite, including various environmental factors, within-human intrinsic factors that affect susceptibility to infection, epidemiological and behavioural determinants of pathogen exposure as well as nutritional and cultural factors, related to food-borne zoonotic diseases. Moreover, the human population density has shown strong correlation with the emerging risks of all major classes of infectious diseases. Thus, it is seemingly clear that human encroachments into species-rich habitats may concomitantly reduce biodiversity and enhance exposure of people to novel microbes with high pathogenicity (Peeri et al., 2020).

RNA viruses are an important group of zoonotic agents that include 180 species, and approximately on average two new species are being discovered yearly. These types of viruses show multiple-host sharing between mammals and birds and thus are considered as the most dreadful agents in zoonotic disease transmission, representing a challenge for global disease control. This group of viruses possess rapid adaptive rates due to an amalgamation of high mutation rates, short generation times, large population sizes, and large mutational selection coefficients linked to anthropogenic-induced rates of interspecies contacts that may lead to a pandemic. Major epidemics outcomes of some RNA viruses include HIV, SARS, Hendra, Nipa, and MERS to name a few (Belouzard et al., 2009)

The undebated role of climate change on several ecological processes has had a huge impact on the ability of many species to adapt to rapid anthropogenic changes causing habitat destruction and space loss. Climate change has shown dramatic effects on all ecological processes including drought in some regions and increased precipitation in others; rise in sea levels leading to increased erosion; increased storms, hurricanes, and tsunamis; and the mass extinctions of many species due to rapid changes in climatic regimes.

However, it is difficult to estimate species numbers and extinction rates since the arrival of humans. A recent report identifies more than 700 threatened mammals and birds that have been adversely affected by climate change (Fuentes et al., 2020). The International Union for Conservation of Nature (IUCN) has estimated extinction rates for several taxa, which include primary threats to 8688 species listed in the IUCN Red Data Book due to overexploitation (72%), agricultural activity (62%), and climate change (19%). Another study by Ceballos et al. (2017) estimated that rate of population loss in terrestrial vertebrates is staggering with 32% of 27,600 vertebrate species showing decrease

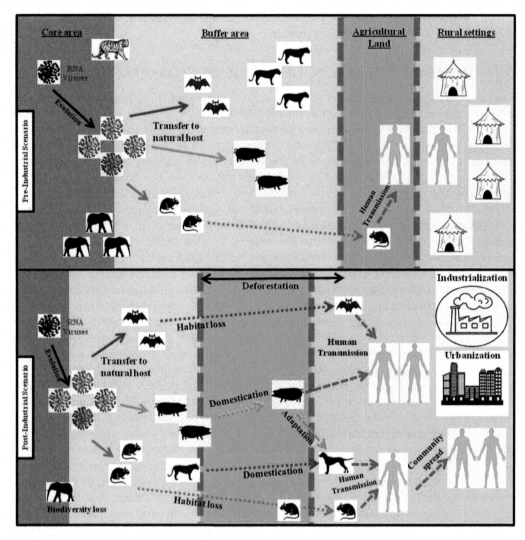

**FIGURE 15.1** Schematic diagram depicting different pathways of cross-species jumping of novel viruses. The process may involve different classes of animals and is dramatically facilitated by various anthropogenic activities like deforestation, bio-diversity loss, and domestication of animals.

in population size and range. The authors refer to these massive declines as a "biological annihilation." Catastrophic and species diminishing effects of such factors have been linked to many species from diverse habitats across the world including nutrient-driven hypoxic dead zones impacting fish and shellfish, Tasmanian devil facial tumors, amphibian chytridiomycosis, neonicotinoids in honeybees, and white nose syndrome of bats. There are many more incidences of species declines and extinctions in which infectious and non-infectious agents have become leading causes. The diseases occurring on human-domestic animal-wildlife interfaces are affected by a multitude of ecological and environmental factors including air and water pollution, anthropogenic-waste mismanagement, urban and industrial microclimates, human encroachment on wildlife habitats, and finally on vector ecology. The effects of anthropogenic changes on the ecosystem health are multifaceted and need transdisciplinary approaches for further assessment and management (Travaglio et al., 2020).

## 15.3   REJUVENATION OF ENVIRONMENT DURING COVID-19 PANDEMIC-RELATED LOCKDOWNS

A large number of recent reports show that COVID-19 pandemic-related lockdowns caused significant improvement in air quality index (AQI) and water quality index (WQI) in many cities across the world. In an attempt to understand the effect of this pandemic on the environment and its related changes upon human anthropogenic activities, we have analysed recent reports on environmental rejuvenation during the lockdown and post-pandemic periods.

### 15.3.1   EFFECT ON AIR QUALITY

As all kinds of industrial, transportation, and other business activities were shut down in the lockdowns, there has been a sudden drop in carbon emission levels and other air pollutants across all countries big and small.

#### 15.3.1.1   Reduced Emissions

Compared to the recorded readings of 2019, the air pollution levels of New York City dropped by almost 50% due to enforced measures to restrict the spread of virus. In China, the use of coal in power plants was reduced by 40% and total emissions show a 25% decrease during initial days of the lockdown in the country with approximately 11% increase in good air quality across many cities of China. Delhi, the national capital of India, has been known as one of most polluted cities in the world in terms of air quality; however, 40–50% improvement in air quality was observed during lockdown (Mahato et al., 2020). In many European countries, like Italy and Spain, satellite images have confirmed drop in nitrogen dioxide ($NO_2$) emission (Ficetola and Rubolini, 2020).

Although economic activities have been stopped and stock markets have fallen, the reduced carbon emission is being seen as a good improvement in carbon and nitrogen footprints. This pandemic has caused major job losses and has threatened the sustenance of millions of people as businesses are either struggling or shutting down to control the spread of virus. Still, it has definitely crafted a decarbonized, maintainable economy that was has been hoped for for decades. Such a pandemic, however, which is taking human lives, should not be seen as a route of bringing about environmental changes. As lockdowns subside, carbon and pollutant emissions will re-appear with no ever-lasting impact on the ecology (McCloskey and Heymann, 2020).

#### 15.3.1.2   Decarbonization

As millions of people across the globe are staying at home due to COVID-19, energy consumption profiles in office buildings are getting unusually disturbed. As a large number of people are working from home nowadays, domestic energy consumption is on the rise. According to prediction data analysis, in the United States the domestic energy consumption has risen by 6% to 8% in 2020 compared to 2019. This pandemic might have a direct effect on reducing the consumption of polluting fuels in power stations and save energy due to drop in demands (Lau et al., 2020). A sudden drop in coal consumption and air pollutants in China is another example of direct effect of the coronavirus on the environment. The number of airborne pollutants like carbon dioxide, carbon monoxide, and nitrous oxides has fallen appreciably. All these incidences reassert the role of COVID-19 in improving the air quality in many regions across the world.

#### 15.3.1.3   Reduced Industrial Waste

Within industrial/manufacturing companies, "working from home" is not a viable option for all assignments. According to a survey conducted by the British Plastics Federation (BPF), COVID-19 lockdown severely impacts manufacturing businesses in the UK with around 80% of the respondents anticipating decline in industrial turnover during two successive quarters, and 98% respondents admitted concern over negative impacts on business operations. The global production of chemical industries is predicted to reduce by 1.2%, which is the worst growth for the sector since the financial crash of 2008. Major chemical manufacturing companies, who were in the process of upscaling their production in China, are reducing their activities, contributing to a slowdown in predicted growth (Allan et al., 2020).

### 15.3.2 EFFECT ON WATER QUALITY

From the beginning, human civilizations were facilitated by river and estuarine systems, which is the reason most human settlements are found on their banks. Beaches are another important natural capital asset found in coastal areas, and provide services (land, sand, recreation, and tourism) that are critical to the survival of coastal communities and have intrinsic values. But due to over exploitation and non-responsible use by people many beaches and rivers in the world have pollution problems (Saadat et al., 2020).

The anthropogenic activities in and around the estuaries (discharges from tourism units, industrial operations, repairing of fishing vessels and trawlers, fish landing, etc.) are leading to poor condition of these estuaries and coastal zones. The agricultural run-offs in the water bodies are escalating the already polluted water bodies, which are leading to eutrophication. The unplanned and uncontrolled proliferation of urban and industrial set-ups leads to increase in discharges and total load of pollutants being dispensed into the water bodies. These discharges may contain various heavy metals (Pb, Cd, etc.) and toxic chemical compounds (DDT, microplastic, etc.) that can bio-accumulate and bio-magnify. The toxic and carcinogenic hydrocarbons such as benzenes, poly-nuclear aromatics, amines, and phenols are present in discharged water from refineries, chemical plants, and offshore drillings (Sharma et al., 2020)

In developing countries like India, which are often threatened with the problem of fecal contamination due to untreated discharges from the sewage canals connected with the rivers (Mukherjee et al., 2020), industrial wastes, factory discharges, agricultural runoffs, and wastes from shrimp farms are also major pollutants discharged. The organic waste and heavy metals in discharge pollute the water bodies and result in increase of Biological Oxygen Demand (BOD) and Chemical Oxygen Demand (COD) in water. Another large source of water pollution is various rituals (such as immersion of mud idols of the Hindu Gods and Goddesses, fruits and flowers and religious offerings of burnt ashes of the dead bodies) and bathing activities performed round the year by thousands of worshippers along river banks, which together with laundry (by the domestic and launderer's community) contaminate the river water with numerous organic and inorganic wastes including organic waste, heavy metals, and hydrocarbons (polycyclic aromatic hydrocarbons) (Agarwal et al., 2020). This results in hundred times' higher levels of fecal coliform bacteria in rivers at specific locations than the official permissible limit of Indian governments. The Ganga Action Plan was started by the Indian government to clean up the river Ganga but failed due to poor environmental planning and technical expertise and a lack of awareness amongst the masses (Table 15.1).

However, the COVID-19 pandemic resulted in positive activity in the environmental sector due to shutdown of the point and non-point sources of pollution (Mitra et al., 2020). To stop the spread of the novel coronavirus, the government of all major countries including India announced lockdowns during February and March 2020. During the lockdown phase all kinds of services (production, transportation, construction, etc.) were suspended except for essential goods and services. Interestingly, these global pandemic-mediated lockdowns improved water quality of river, estuaries, and beaches to a great extent. For example, beaches like those of Barcelona (Spain), Acapulco (Mexico), or Salinas (Ecuador) now look cleaner and with crystal clear waters (Rosenbloom et al., 2020).

The COVID-19 lockdowns helped in eco-restoration of environmental quality throughout the world including in big cities in India such as Kolkata (Mitra et al., 2020). The complete cease

**TABLE 15.1**

**List of Water Parameters/Pollutants, Sources of Pollutants, and Their Status Before and During Lockdown**

| Parameter/Pollutant | Source | Pre-Lockdown | Lockdown | References |
|---|---|---|---|---|
| Dissolved Oxygen | Atmospheric $O_2$ | Gradual Decrease | Increase of 10–30% | (Dhar et al., 2020) |
| Heavy Metals (Zn, Cu, Pb) | Industrial Effluents | Above Permissible limits | Decrease of 5–10% | (Agarwal et al., 2020) |
| Total Colioform (TC) | Sewage Wastes | High Matrix | Drastic Decrease | (Mukherjee et al., 2020) |
| Ichthyoplankton | Natural Indicator of Oil Spill | Adverse Impact | Restoration | (Mitra et al., 2020) |
| Nitrate and Phosphates | Sewage Releases | Large Amounts | Sharpe Decline | (SenGupta et al., 2020) |
| Acidification (pH) | Enhanced Atmospheric $CO_2$ | Acidified (<7) | Steady Increase in pH | (Dutta et al., 2020) |

of industrial operations, traffic movements, and tourism along with lack of community bathing activities and various religious rituals along the banks of the rivers during lockdowns upgraded the estuarine water quality as revealed by the hike in DO values and sharp decline in the total coliform, which has several positive implications particularly in the domain of sustaining the water ecosystem. The results of the present study shed light on the importance of reduction in human intervention and drastic improvement in the microbiological quality of the river environments (ganges in this case).

The lockdown period witnessed general improvement in water quality in Yamuna, a major river in India. During lockdown, a complete absence of effluent discharge from all 23 sources and strict restrictions on other activities caused improvement in all quality parameter like pH, DO, EC, BOD, and COD, compared to the pre-lockdown phase in the Yamuna River. Various action plans and efforts were made to clean Yamuna where a huge amount of money was invested but never seen as satisfied result as seen in lockdown phase, during this phase Yamuna gets revived (Arif et al., 2020).

Several recent studies related to eutrophication, dissolved zinc contents, pH, oil, planktonic populations, BOD and Dissolved Oxygen (DO), and nitrite and phosphate (Table 15.1) quantifications and comparisons before and during lockdown conditions have revealed that COVID-19 lockdowns improved natural water resource quality without any major investment in treatment processes (Dhar et al., 2020).

### 15.3.3 EFFECT ON GLOBAL MOVEMENT RATE

#### 15.3.3.1 Educational Travels

The education sector is highly affected by COVID-19 from preschool to tertiary level of education. Universities and research institutes have moved online, exams have been postponed, there have been delays in award of degrees and certificates, huge declines in new jobs and interviews, and major research projects are on hold or have been cancelled. Many researchers shifted to digital work to continuing employ students and technicians by working remotely on data analysis, digitizing paper records, and coding interview transcripts. It has also been possible to conduct interviews, training sessions, other activities outdoors (Corbera et al., 2020). But such delays in specialized trainings and career growth, in combination with lurking economic and psychological pressures of the pandemic, may cause a large cohort of students to leave their field of interests and choose other career options that offer more stability.

### 15.3.3.2   Research Activities

Activities of university laboratories and other research facilities were closed, thereby putting a pause on major experiments and new research. The impact on field research has been impacted more harshly due to travel restrictions and accessibility issues. The threat of disease transmission has abated field-based social research and focus-group interviews. The lockdown-induced global economic recession has led to cuts in funds for research grants in various research projects and other public-funded programs. The Global Environment Facility (GEF) and Inter-American development bank (IDB) have started collecting data for risk analysis of projects funded by them, which were affected during lockdown (Staniscuaski et al., 2020).

### 15.3.3.3   Tourism, Hospitality, and Aviation

#### 15.3.3.3.1   TOURISM

Tourism has been one of the most harshly impacted sectors by the COVID-19 outbreak, due to complete halt in both supply and demand. It has been predicted that approximately 50 million jobs in the tour and travel sector may be lost or be at risk. Similarly, in Europe, the alliance of European Tourism Manifesto, which comprises more than 50 European public and private organisations of tour and travel sector, had pressed upon the urgent need to launch the European Unemployment Reinsurance Scheme. According to a report, international visitors to Vietnam were reduced by more than half during the first quartile of 2020 as compared to the first quartile of 2019. Similarly, the Philippines is projecting a 0.3–0.7% slowdown in the yearly GDP of the country. The United States and United Kingdom have put restrictions on all non-essential travel and sealed their borders for foreign tourists in turn causing accelerated disruption of their economies (Brounder et al., 2020).

#### 15.3.3.3.2   HOSPITALITY

The hospitality sector, along with the travel industry, has also been hit hard, with hourly workers facing potentially devastating hardships. The occupancy of rooms and revenue generated per available rooms in the United States and China have tanked to unusually low levels in 2020. Hotel occupancy in European countries including Germany and Italy has decreased by over 36%. Thus, COVID-19 pandemic-induced lockdowns led to international disruption of the hospitality industry across all nations, bringing a slump in the hotel market (Karim et al., 2020).

#### 15.3.3.3.3   AVIATION

The unforeseen conditions of social distancing, restricted travelling, and closure of international boarders caused sudden cancelations in travels and a huge drop in demand for the travel sector. The US barred all foreign nationals from Iran, China, and many EU countries from entry in the grid. Similarly, the Foreign Office of UK also advised its nationals for essential international travel only. There was a suspension of all non-essential travel for thirty days from outside Europe, following which the Malaysian airport has reported a 30% decline in international passengers (Sobieralski et al., 2020).

### 15.3.3.4   Other Communications and Networking

Web-based (online) conferencing technology has provided an effective way for small- to medium-sized gatherings, where people may interact and exchange ideas, but these virtual meetings are not an easy replacement for the in-person gatherings where networking and interactions develop. A cancelled conference may be of minor concern for established professionals; however, it is a huge loss for beginners like graduate students and early career researchers (Manderson & Levine et al., 2020).

## 15.4   ENVIRONMENTAL MANAGEMENT AND SDG (SUSTAINABLE DEVELOPMENT GOAL): CHANGE IN PERSPECTIVES, POST-LOCKDOWN

The COVID-19 crisis is global and unpredicted and requires coordinated responses by policymakers, businesses, and broader society. Failure of the human race to anticipate the approach of a pandemic of this scale has created havoc. Thus, it is important to consider some other scenarios for the future looking at the ongoing conditions of countries in this COVID-19 pandemic. The pandemic-induced impacts should be understood fully and used to develop a better and different future society.

The recovery funds should be used to stimulate innovation for the low-carbon energy transition, promoting new infrastructure, business models, and industrial capacity in the field of renewable energy technology (energy storage, electric vehicles, and charging stations) through tax credits and other measures. During this phase we are leveraged to accelerate the phase-out of coal-fired power, which is already in the climate action plans of several countries (Nicola et al., 2020).

There is an urgent need for research and development on building effective methods to provide early and accurate detection of such diseases so that morbidity and mortality conditions of human race can be improved. Robust technologies like Internet of Things (IoT) and Big Data Analysis (DBA) can form the basis of future research for mapping and alarming the spread of infection (Tehsin, 2017; Rifai et al., 2020).

Another impact on economy and environment is the surge in plastic usage during lockdown, which has not been completely analysed until now. The large amount of plastic waste generated globally may surpass the existing disposal or treatment facilities, posing risk of secondary issues due to improper waste management (Klemes et al., 2020). The waste management planning needs supports from the latest decision-making tools, mobilised/automated (e.g., remote-controlled robots) treatment and collection design, treatment approaches, scalability, infrastructure, logistics, safety, and regulatory aspects linked to bio-disaster response. There is need to rethink the strategies for minimising impacts of plastic and making use of its merits at the same time (Huang et al., 2020). Serious considerations about society and recycling of plastics should be the focus of considered authorities. Social factors are the most neglected portions of human attempts towards achieving sustainability (Singh et al., 2020). Countries also have to invest more in health care – both management and facilities. In the current scenario, the new normal is following new social behaviour "social distancing," "wearing masks," "maintaining hygiene," etc. (Gutiennez et al., 2019). Future research should be based on mapping the spread of infection by using implications of IoTtechnologies (Dev et al., 2020). Sustainable recovery should be used for climate-positive changes particularly in fossil-intensive companies and aviation (Wilkinson et al., 2018).

While SDGs were planned to be a progression of the UN's previous Millennium Development Goals (MDGs) (United Nations, MDGs and Beyond 2015), it remains to be seen how they will tackle this situation and whether they evolve to stand on their own in a post-COVID-19 world. The novel coronavirus impacted all ongoing SDGs in significant ways and challenged the global commitment to rethink and replanned toward vision 2030. The COVID-19 challenges SDG-1 as it likely increased poverty especially in impoverished nations that do not have adequate resources to respond against the virus and protect their populations. The acceleration is needed in development of vaccines and therapies against any pandemic mitigation to sustain good health and well-being (SDG-3). Investments in medical and public health infrastructure (SDG-9) are needed as we have witnessed our medical infrastructure failure during this pandemic. For better response against global challenges such as COVID-19 we need to align and synchronise with other nations around the globe and their health systems. The workforce must be retrained and rethought to ensure decent employment and economic growth (SDG-8) in a post-COVID-19 world for adaptable, competitive, and flexible work environments. The retraining and rethinking needed nations workforce to ensure decent work and economic growth (SDG-8) in a post-COVID-19 world for versatile, agile, and competitive under elastic work environments. Working places should be rearrange to maintain social distancing if needed in these types of pandemics. Partnerships (SGD-17) with other nations to renew their commitment and support for research on sustainable development especially to global health and wellbeing should be developed. To ensure sustainability of the urban ecosystem we have to ensure a healthy urban environment and its people to achieve new research approaches (Rifai et al., 2020).

## 15.5   CONCLUSIONS AND FUTURE PROSPECTS

Evidence is showing that loss of habitat due to deforestation and environmental degradation and consumption of meat of virus-infected wildlife and domesticated animals may lead to transmission of animal viruses to human beings. Climate change and disasters, rapid decline in the population of natural hosts, chemical pollutants, and increasing irradiation may cause rapid mutations in pathogenic strains that can turn out be more or less virulent than their parents. We are living in a world with increasing threats to deadly communicable diseases in an era of climate change and environmental degradation. The richness of air, water, and biodiversity are associated with anthropogenic activities, heavy transportation using fossil fuel-based vehicles and clustering of people in cities, metros, and markets. The COVID-19 caused global lockdowns but gave a message to minimize the clustering and use of non-renewable energy sources to mitigate the negative impacts of climate change, pollution, and environmental degradation. Although several recent studies suggest COVID-19-related reduction of environmental impacts it is still too early to understand and comment upon the final picture of environmental effects. There is still uncertainty in the economic recovery path, and consumers may change their needs and demands.

Green investments in nature-based solutions like net-negative emission, agriculture, and infrastructure for resilience should be made in industrial infrastructures and urban developments to mitigate anticipated future climate damage.

Efficient internet connectivity to the down human settlements and strengthening of solar cells and biomass-based components of the future planning and business development through the world India has a benefit of high-intensity solar radiation throughout the year other than about a short winter of nearly two months. The sugar mills are getting equipped to produce biofuel and biogas as a product of sugancane biomass, which will be a good income source for sugarcane growers.

India has played a pivotal role in the post-Covid-19 world, and it needs to get engaged in global discussions and efforts and at the same time try to develop methods and innovations to lead the world. India must make efforts towards achieving permanent and sustainable solutions for international problems and their solutions. To work towards a safer and greener planet, every single step considering the complexity of various issues is an imperative goal of humankind.

## ACKNOWLEDGMENTS

AVA would like to acknowledge Department of Science & Technology, New Delhi (India) for financial support through Science, Technology, Innovation PostDoctoral Fellowship. PK is thankful to DBT for financial support through SRF.

## CONFLICT OF INTERESTS

The authors declare no conflict of interest.

## REFERENCES

Agarwal S, Pramanick P, Mitra A (2020). Alteration of dissolved Zinc concentration during COVID-19 lockdown phase in coastal West Bengal. *NUJS J Regul Stud* (Special Issue).

Aguirre AA (2017). Changing patterns of emerging zoonotic diseases in wildlife, domestic animals, and humans linked to biodiversity loss and globalization. *ILAR J, 58*(3), 315–318.

Ali I, Alharbi OM (2020). COVID-19: Disease, management, treatment, and social impact. *Sci Total Environ*, 138861.

Allan J et al. (2020). A net-zero emissions economic recovery from COVID-19. *COP26 Universities Network Briefing.*

Arif M, Kumar R, Parveen S (2020). Reduction in water pollution in Yamuna river due to lockdown under COVID-19 pandemic.

Ather A, Patel B, Ruparel NB, Diogenes A, Hargreaves KM (2020). Coronavirus disease 19 (COVID-19): implications for clinical dental care. *J Endod.*

Belouzard S, Chu VC, Whittaker GR (2009). Activation of the SARS coronavirus spike protein via sequential proteolytic cleavage at two distinct sites. *PNAS, 106*(14), 5871–5876.

Brouder P, et al. (2020). Reflections and discussions: Tourism matters in the new normal post COVID-19. *Tour Geogr*, 1–12.

Ceballos G, Ehrlich PR, Dirzo R (2017). Biological annihilation via the ongoing sixth mass extinction signaled by vertebrate population losses and declines. *PNAS, 114*(30), E6089–E6096.

Corbera E, Anguelovski I, Honey-Rosés J, Ruiz-Mallén I (2020). Academia in the time of COVID-19: Towards an ethics of care. *Plan Theory Pract*, 1–9.

Dev SM, Sengupta R (2020). Covid-19: Impact on the Indian economy. *IGIDR,* Mumbai, April.

Dhar I, Biswas S, Mitra A, Pramanick P, Mitra A (2020). COVID-19 lockdown phase: A boon for the River Ganga water quality along the city of Kolkata. *NUJS J Regul Stud* (Special Issue).

Dutta P, Pramanick P, Biswas P, Zaman S, Mitra A (2020). Reversing the phenomenon of acidification in the River Ganges: A ground zero observation. *NUJS J Regul Stud* (Special Issue).

Ficetola GF, Rubolini D (2020). Climate affects global patterns of COVID-19 early outbreak dynamics. *medRxiv.*

Fuentes R, Galeotti M, Lanza A, Manzano B (2020). COVID-19 and climate change: A tale of two global problems. Available at SSRN 3604140.

Gutierrez B, Escalera-Zamudio M, Pybus OG (2019). Parallel molecular evolution and adaptation in viruses. Curr Opin Virol, *34*, 90–96.

Huang C et al. (2020). Clinical features of patients infected with 2019 novel coronavirus in Wuhan, China. *The Lancet, 395*(10223), 497–506.

Jaimes JA, André NM, Chappie JS, Millet JK,Whittaker GR (2020). Phylogenetic analysis and structural modeling of SARS-CoV-2 spike protein reveals an evolutionary distinct and proteolytically-sensitive activation loop. *J Mol Biol.*

Karim W, Haque A, Anis Z, Ulfy MA (2020). The movement control order (mco) for covid-19 crisis and its impact on tourism and hospitality sector in malaysia. *ITHJ, 3*(2), 1–7.

Klemeš JJ, Van Fan Y, Tan RR, Jiang P (2020). Minimising the present and future plastic waste, energy and environmental footprints related to COVID-19. *Renewable Sustainable Energy Rev., 127*, 109883.

Kooraki S, Hosseiny M, Myers L, Gholamrezanezhad A (2020). Coronavirus (COVID-19) outbreak: What the department of radiology should know. *J Am Coll Radiol.*

Kucharski AJ, Russell TW, Diamond C, Liu Y, Edmunds J, Funk S, ... & Davies N (2020). Early dynamics of transmission and control of COVID-19: A mathematical modelling study. *Lancet Infect Dis.*

Lau ET, Yang Q, Stokes L, Taylor GA, Forbes AB, Clarkson P, ... & Livina VN (2015). Carbon savings in the UK demand side response programmes. *Appl Energy, 159*, 478–489.

Li JY, You Z, Wang Q, Zhou ZJ, Qiu Y, Luo R, Ge XY (2020). The epidemic of 2019-novel-coronavirus (2019-nCoV) pneumonia and insights for emerging infectious diseases in the future. *Microbes Infect, 22*(2), 80–85.

Mahato S, Pal S, Ghosh KG (2020). Effect of lockdown amid COVID-19 pandemic on air quality of the megacity Delhi, India. *Sci Total Environ*, 139086.

Manderson L, Levine S (2020). *COVID-19, risk, fear, and fall-out.*

McCloskey B, Heymann DL (2020). SARS to novel coronavirus–old lessons and new lessons. *J Infect Dis Epidemiol, 148.*

Messner CB, Demichev V, Wendisch D, Michalick L, White M, Freiwald A, ... & Ludwig D (2020). Ultra-high-throughput clinical proteomics reveals classifiers of COVID-19 infection. *Cell Syst.*

Mitra A, Pramanick P, Zaman S, Mitra A (2020). Impact of COVID-19 lockdown on the *Ichthyoplankton* community in and around Haldia Port-cum-Industrial complex.

Mukherjee P, Pramanick P, Zaman S, Mitra A (2020). Eco-restoration of River Ganga water quality during COVID-19 lockdown period using Total Coliform (TC) as proxy. *NUJS J Regul Stud* (Special Issue).

Nicola M et al. R (2020). The socio-economic implications of the coronavirus pandemic (COVID-19): A review. *Int J Surg (London, England), 78*, 185.

Olivero J et al. (2017). Recent loss of closed forests is associated with Ebola virus disease outbreaks. *Sci Rep*, *7*(1), 1–9.

Peeri NC et al. (2020). The SARS, MERS and novel coronavirus (COVID-19) epidemics, the newest and biggest global health threats: what lessons have we learned? *Int J Epidemiol*.

Rifai HS (2020). The sustainable development goals in a bioremediation journal context: what a difference a year makes in a post COVID-19 world!

Rodriguez-Morales AJ et al. (2020). History is repeating itself: Probable zoonotic spillover as the cause of the 2019 novel Coronavirus Epidemic. *Infez Med*, *28*(1), 3–5.

Rosenbloom D, Markard J (2020). A COVID-19 recovery for climate. 447–447.

Rulli MC, Santini M, Hayman DT, D'Odorico P (2017). The nexus between forest fragmentation in Africa and Ebola virus disease outbreaks. *Scientific Reports*, *7*, 41613.

Saadat S, Rawtani D, Hussain CM (2020). Environmental perspective of COVID-19. *Sci Total Environ*, 138870.

Sengupta T, Pramanick P, Mitra A (2020). Nutrient load in the River Ganges during the COVID-19 lockdown phase: A Ground Zero observation. NUJS Journal of Regulatory Studies Special Issue.

Sharma R, Hossain MM (2020). Household air pollution and COVID-19 risk in India: A potential concern.

Singh SK (2020). COVID-19: A master stroke of Nature. *AIMS Public Health*, *7*(2), 393.

Sobieralski JB (2020). COVID-19 and airline employment: Insights from historical uncertainty shocks to the industry. *TRIP*, 100123.

Staniscuaski F et al. (2020). Impact of COVID-19 on academic mothers. *368(6492), 724*

Tehsin RH (2017). Suggestions to lessen the man animal conflict. *Indian For*, *143*(10), 1093–1093.

Travaglio M, Yu Y, Popovic R, Leal NS, Martins LM (2020). Links between air pollution and COVID-19 in England. *medRxiv*.

Venter ZS, Aunan K, Chowdhury S, Lelieveld J (2020). COVID-19 lockdowns cause global air pollution declines with implications for public health risk. *medRxiv*.

Wilkinson DA, Marshall JC, French NP, Hayman DT (2018). Habitat fragmentation, biodiversity loss and the risk of novel infectious disease emergence. *J. R. Soc. Interface*, *15*(149), 20180403.

Zhou F et al. (2020). Clinical course and risk factors for mortality of adult in patients with COVID-19 in Wuhan, China: a retrospective cohort study. *The Lancet*.

# 16 COVID-19 Pandemic and Environmental Pollution

## Opportunity to Revisit Environmental Strategies for Public Health Benefits

*Suman Mor,*[1*] *Akshi Goyal,*[1] *Tanbir Singh,*[1] *Sahil Mor,*[2]
*R. C. Sobti,*[3] *and Khaiwal Ravindra*[4]

[1]Department of Environment Studies, Panjab University,
Chandigarh, India
[2]Department of Environmental Science and Engineering, Guru
Jambeshwar University of School and Technology, Hisar, Haryana, India
[3]Department of Biotechnology, Panjab University, Chandigarh, India
[4] Department of Community Medicine and School of Public Health,
PGIMER, Chandigarh, India
*sumanmor@yahoo.com

## CONTENTS

## 16.1  INTRODUCTION

On December 31, 2019, cases of pneumonia with an unknown etiology were reported in the city of Wuhan, Hubei province in China. All cases were found to be linked with the Wuhan seafood market, which trades a variety of live animal species (Chinazzi et al., 2020). The pathogen responsible for causing severe pneumonia-like symptoms was recognized as Severe Acute Respiratory Syndrome Coronavirus 2 (SARS-CoV-2). Later, the World Health Organization (WHO) named the disease COVID-19 (Coronaviridae Study Group of the International, 2020). The coronavirus is believed to be originated from bats and transferred to humans. The virus is known to be transmitted on inhalation or by coming into contact with infected droplets produced while sneezing, coughing, and talking by symptomatic or asymptomatic people (Singhal, 2020). The lifespan of droplets varies from surface to surface, which have a tendency to travel through air over some distance. The incubation period of the disease varies between 2 and 14 days.

The signs and symptoms include fever, sore throat, cough, loss of sense of smell, taste, breathlessness, and fatigue. But the clinical symptoms vary from asymptomatic patients to acute respiratory distress syndrome (ARDS). Globally, more than 23.12 million cases of coronavirus have been reported with around 803,253 deaths all over the world, whereas more than 15.7 million cases have been recovered spontaneously (WHO Report, 2022). However, in some cases (elderly and those with multiple co-morbidities) the virus leads to complications like organ failure, cardiovascular disease, gastric diseases, pneumonia, and ARDS. The fatality rate associated with COVID-19 ranges from 2 to 4%. The virus is reported to be destroyed by using common disinfectants (e.g., hydrogen peroxide) and alcohol-based sanitizers.

On January 30, 2020, the WHO declared the outbreak of COVID-19 as a public health emergency of international concern putting various countries at high risk. According to the WHO, the spread can be contained via early detection, isolation, treatment, and implementing robust rules towards COVID-19 (Sohrabi et al., 2020). After declaration of this outbreak as a pandemic, many countries in the world completely suspended transport across the international borders to prevent COVID-19 transmission. Further, to restrict the spread of coronavirus the unprecedented lockdown was implemented across several countries and territories ranging from weeks to months.

The term lockdown refers to to "mass quarantine" and typically refers to "stay-at-home" ordinances given by national or regional government/authorities for limiting movement/activities of people in designated areas for specific time periods. It is mainly used to limit outbreaks, mandating citizens to stay at home, except for carrying out essential activities such as groceries, chemists, pharmacies, water, electricity, sanitation services, police services, health services, medical emergency, etc. (Lippi et al., 2020). It is estimated that approximately one-third of world's population (i.e., 1.3 billion in India, 50–60 million in China, 280 million in Europe, and around 150 million in the United States) have been subjected to comply with restrictive measures under lockdown (*The Brussels Time*, 2020).

Lockdown orders have serious effects on the world's economy, environment, and society, especially when it is extended for months. Although COVID-19 lockdowns have had many repercussions on the world's economy and underprivileged people in particular, it has also had positive impacts on the environment. In this chapter, the impact of lockdown and restricted human activities on environment indices such as air and water noticed are discussed (Zambrano-Monserrate et al., 2020). Moreover, its impact on biodiversity, behavioural change in humans, and lessons learned during this pandemic for environmental protection and public health benefits are discussed.

## 16.2  COVID-19 LOCKDOWN AND ITS IMPACT ON ENVIRONMENTAL POLLUTION

The outbreak of COVID-19 forced many countries to undergo partial and full lockdown ranging from a few weeks to months. All activities including tourism, industrial, vehicular movement along with cultural, social, religious, and mass gatherings were prohibited (Chakraborty and Maity, 2020). In response to limited human movement and activities, substantial reduction in levels of air pollutants such as $PM_{2.5}$, $PM_{10}$, $CO_2$ and $NO_2$ and improvement in Air Quality Index (AQI) was noticed (Cadotte, 2020). Since industrial and vehicular activities are the major source of $NO_2$ emission in the atmosphere and both activities came to a halt for 30–90 days in many countries, during this time substantial decrease in the level of $NO_2$ emissions in many parts of the world were reported by the Centre for Research on Energy and Clean Air (CREA). Moreover, radial drop in $NO_2$ emissions mainly across China, Spain, Germany, and Italy were reported by the European Space Agency (ESA) through imagery (ESA 2020a, 2020b).

Consumption of fossil fuels is considered as the main source of $CO_2$ emissions from industries and factories and their shut down led to drastic reduction in $CO_2$ emissions (Isaifan, 2020). China contributes the largest fraction of carbon emissions and a reduction in $CO_2$ emissions of approximately 18% during the period of February and March 2020 was seen (Chen et al., 2020). Whereas reduction of 390 metric tonnes and 250 metric tonnes of carbon emissions were reported in Europe and UK, respectively (Paital, 2020). In many cities, people were experiencing clear blue skies for the first time in a long while. As the pollution level dropped in India, the Dhauladhar range of Himachal was clearly visible from Jalandhar (Punjab), which is approximately 200 km far away. Along with the drop in the level of air pollution, multiple ecological events have also been observed in surface water bodies including rivers in many cities all around the world (Kohli, 2020). As per the ESA, the Venice canal of Italy appeared to be very clear and small fishes begun to inhabitant the waterways once again. Similarly, due to shutdown of various industrial activities and related industrial discharge in water bodies, the absence of ritual activities in waterbodies, and other human activities near water bodies, River Yamuna and Ganga become pollution free and appeared much cleaner than before (Wright, 2020).

Human activities are reported as a leading cause of noise pollution from residential, commercial (vehicular traffic, loud speakers), and industrial sources. Remarkably, noise pollution contributes to various health issues including elevated blood pressure, sleep interference, mental stress, hearing loss, and increased heart rate. Various noises from different sectors also have a huge negative impact on birds, fishes, snakes, and mammals as they have different sensitivity ranges to noise (Schuster, 2020). During lockdown, as all the activities were standing still, the noise footprint was found to be reduced in many countries including India and the UK (Zambrano-Monserrate et al., 2020). With less human movement, the planet literally calmed and seismologists reported lower vibrations from "cultural noise" than before the pandemic (Watts, 2020), which can be considered as another positive sign for self-recovery of nature. Hence, significant positive impacts on environment quality were observed.

Despite of these positive impacts, during the COVID-19 pandemic an adverse impact on waste management sector has also been reported. A study conducted in Morocco assessed the influence of items purchased and consumed during lockdown period and found that quantity of organic waste decreased in comparison to inorganic waste whereas 87% of population was found to mix their bio-medical waste with household waste, a leading factor that can lead to spread of coronavirus (Ouhsine et al., 2020). During the pandemic extensive use of various protective gears like gloves, face masks, plastic shields, sanitizer bottles, etc., increased, which resulted in the generation of a huge amount of bio-medical waste. Increased rate of bio-medical waste has also been observed during COVID-19 in Wuhan, where an average of 240 metric tonnes/day was produced compared to the previous average of <50 tonnes. Further, in the United States, to avoid spread of the virus, recycling programs have been suspended in some cities. And in some European nations sustainable waste management has been restricted to curb the coronavirus. For example, Italy prohibited sorting residential waste from infected population areas (Zambrano-Monserrate et al., 2020). All these activities are considered responsible for increasing both domestic and biomedical waste, which poses various environmental issues. This mixed waste remains a major cause of concern and can lead to serious environment and public health threats. Therefore, urgent attention is needed to address the challenges due to the large amount of biomedical waste during this pandemic period, which gets mixed with normal household waste and poses a threat to public health.

### 16.2.1 Nature and Biodiversity

Globally, many efforts are being made by various national and international groups in order to conserve and protect our natural environment. While the ecosystem is shared by all living organisms it is dominated by humans. Therefore, restricting human activities can help to restore our ecosystem. The pandemic-induced lockdowns to "flatten the coronavirus curve" pushed a large number of people of the world to stay at home by shutting down schools and businesses, transportation, industrial activities, etc. (Helm, 2020). This period provided an opportunity for nature to self-regenerate. People also noticed this self-retrieval of nature during lockdown and spontaneous changes in environment quality (Saadat et al., 2020). During lockdown the restricted presence of human beings provided an opportunity for wildlife to roam freely and many were seen moving within various cities, roads, and even in human habitation across the world. For example, a coyote was seen near the Golden Gate Bridge in San Francisco, US, while deer were found to grazing near Washington, boars were found roaming in the city and roads of Barcelona, Spain, and whales appeared in Vancouver after decades. Many endangered animals were also seen in various parts of the world. Malabar civet appeared at zebra crossing in Kozhikode, India,endangered Olive Ridleysreturned to dig their nests as there is no human interference due to lockdown in Odisha's Rushikulya rookery and Gahirmatha Beach, India, endangered dolphin spotted in Mumbai, India. All these incidences suggest that with little or no interference of humans, nature has the capacity to regenerate itself. This pandemic has displayed various contrasting consequences on human civilization as on the one hand it restricted human activity, but on other hand it supported the environment leading to a positive impact on natural biodiversity.

## 16.3  LOCKDOWN AND ENVIRONMENTAL POLLUTION REDUCTION: LEARNINGS FOR CONTROL STRATEGIES

As mentioned earlier COVID-19 resulted in lockdowns in many countries, which in turn resulted in less air pollution and greenhouse gas emissions and improved air quality (Sharma et al., 2020). Once the period of lockdown and partial lockdown was over, the global economic activities started to revert back to pre-lockdown period and hence started to contribute to environment pollution. Therefore, declining levels of greenhouse gases for short periods of time cannot be considered as a sustainable way to clean the environment (Zambrano-Monserrate et al., 2020).

COVID-19 has not only given us the opportunity to restore the environment, it has also presented a strategic opportunity for policymakers to transit towards a more environment sustainable post-COVID world based on what has been seen during lockdown (Rosenbloom and Markard, 2020).

### 16.3.1  Ecological Solutions

Industrialization, urbanization, and uncontrolled growth of the human population have led to deforestation. Deforestation is considered as a key driver of zoonotic disease emergence (Poudel, 2020). As per the World Livestock Report (2013) of the United Nation's Food and Agriculture Organization (FAO), 60–70% transferable diseases emerging in humans are from wild animals (Arora and Mishra, 2020; Yang et al., 2020).

COVID-19 emerged from the Hunan Seafood Market of Wuhan, China, which is engaged in trading of a variety of live animals including bats, pangolin, snakes, etc. Millions of dollars have been spent on developing treatment and medicine for the disease, but the primary tool for prevention has been long-forgotten and neglected such as afforestation and preserving of wildlife habitats (Afelt et al., 2018; Olivero et al., 2017).

Therefore, there is an urgent need to understand the importance of wildlife habitats and the significance of forests. Many scientists and researchers have also pointed to the need for different countries to put a permanent ban on wildlife trades and markets (Wang et al., 2020; Yang et al., 2020). Strict measures towards wildlife trade regulations and extensive measures to protect natural environments are needed, with a focus on long-term ecological solutions to avoid outbreak of zoonotic diseases. Some recommendations and strategic approaches to avoiding emerging zoonotic diseases are listed in Table 16.1. Another way to tackle these pandemics includes achieving targets

**TABLE 16.1**

**Recommendations and Strategic Approaches to Avoid Emerging Zoonotic Diseases**

- One World, One Health Approach must be followed to understand and deliver the inherent challenges of complex zoonotic diseases.
- Global efforts to strengthen the Disease Surveillance system to early detect zoonotic outbreaks and identify human population at higher risk to better assess the risk and challenges.
- Seamless collaboration required between animal health and public health agencies at state, national and international level.
- Engagement of non-governmental organizations to provide wide geographic reach and expertise for better surveillance.
- Better analysis of the magnitude and distribution level of zoonotic diseases with strengthen of laboratory network and help of researchers.
- Prevention and control of zoonotic diseases based on synchronization, collaborative fieldwork between multidisciplinary teams to conduct investigation of new and emerging zoonoses by use of diagnostic tools, information technology, communication and advance epidemiological and molecular biology methods.
- Behavioural modification campaigns to educate people with help prevention posters, interactive education sessions and graphic descriptions to reduce the risk of zoonotic diseases.
- Outreach activities to increase awareness and promote behavioural change towards environment and biodiversity.
- Provide information to professionals working in close proximity of animals such as hunters to take precautionary measure (proper hygiene, safety procedures, and personal protective equipment) to reduce the risk for contracting zoonotic diseases.
- Reducing the risk of illegal wildlife trade by increasing policy intervention.

of UN environment sustainability goals (SGDs) to secure livelihoods and protect nature (Arora and Mishra, 2020; Booth et al., 2020).

### 16.3.2 A Step towards Sustainability

Although emissions from different sectors contributing GHGs decreased during lockdown, which in turn lead to development of confidence in regulatory bodies that significant improvement in environment could be achieved, if strict execution of control plans and strategies are not implemented through wise short-term and long-term structural changes adopted by different countries through policy ratification and environmental commitments the world will go back to the way it was pre-Covid. Reducing environmental pollution is a cumulative task and requires community participation, technological advancement, policies and behaviour change towards the environment stimulated by government regulations, and making mandatory changes in transportation activity, investing in cleaner technology and a shift towards renewable energy (Saadat et al., 2020). A strategy to disrupt the promotion and acceleration of carbon-intensive industries and simultaneously accelerating the phasing out of coal-powered plants, already a part of the Climate Action Plan (Rosenbloom and Markard, 2020), need to be implemented. Moreover, more holistic approaches with integrated frameworks should be developed to increase or improve the relationship between humans and the environment.

## 16.4 ENVIRONMENTAL POLLUTION REDUCTION AND PUBLIC HEALTH BENEFITS

Global environmental pollution including air pollution, greenhouse emissions, acid deposition, water pollution, and waste generation is caused by urbanization and industrialization (Briggs et al., 2003). The WHO has identified air pollution as a major environmental health risk factor. Air pollution is found to be associated with increasing mortality and morbidity of respiratory and circulatory diseases. According to the WHO, 4.2 million premature deaths annually are due to outdoor air pollution (WHO, 2018). Diseases like asthma, stroke, bronchitis, respiratory allergies, heart and lung diseases are also attributed to air pollution (Saleh et al., 2020). Besides these chronic and short-term effects other factors such as psychological and behavioural changes are also reported due to air pollution.

Due to lockdown some unexpected consequences were seen in terms of lowering levels of air pollution and improved air quality in most parts of the world including India, China, Europe, and Italy (Chen et al., 2020; Gautam et al., 2020). During a few months, China experienced reduction of $NO_2$ and $CO_2$ emissions by 30% and 25%, respectively (NASA, 2020), whereas $CO_2$ emissions were reduced by 6% worldwide (Carbon Brief, 2020).

In India, a total of 91 cities were under the "Good" and "Satisfactory" category and no city was under "Poor" in terms of AQI on March 29, 2020 (SAFAR-India 2020; Anjum, 2020). Meanwhile, the National Aeronautics and Space Administration (NASA) and ESA released images showing significant changes in air quality during lockdown over China and India (ESA, 2020a). Multi-satellite images showed declines in levels of $NO_2$ over India (Biswal et al., 2021).

Decreases in the level of air pollution noticed during the lockdown could have been linked with decreased mortality rate due to air pollution in the absence of the COVID-19 pandemic. Moreover, due to reduction in air pollution positive health effects could have been studied (Dutheil et al., 2020). Apart from this, noise pollution is also associated with many physiological and mental health issues such as sleep interference, internal stress, high blood pressure, stroke, and increased heart rate. The remarkable decrease in air, noise pollution, and water pollution during lockdown (Mandal and Pal, 2020) may have directly benefitted human health. However, comprehensive studies are needed to study the impact of improved environmental quality on health. The lockdowns provided an excellent opportunity to rethink the anthropogenic interventions made by humans on the environment.

## 16.5 HUMAN BEHAVIOURAL CHANGE AFTER COVID-19 PANDEMIC: LESSONS FOR ENVIRONMENTAL PROTECTION

The COVID-19 pandemic significantly affected life all over the world. Within a few months, thousands of people died and millions were infected with the SARS-CoV-2 virus. During lockdowns, the majority of people stayed home, working and learning remotely, following social distancing norms and maintaining cleanliness around them along with following other advisories issued by the government. This resulted in mutual agreement among communities, that a deteriorating environment can impact human life and nature can reclaim itself if is left undisturbed. Moreover, the pandemic provided time to pause and think about the environment and good hygiene practices for better health and to avoid the spread of COVID-19. This pandemic has changed people's mindset, their eating habits and lifestyles (Ling and Ho, 2020). It has been reported that people increased consumption of fruits and vegetables and avoided meat products in order to strengthen or boost their immune system to better fight against COVID-19 (Ouhsine et al., 2020).

Lockdown also led to people becoming more aware of their physical health as well as that of others and developing habits such as wearing facial masks, social distancing, and becoming knowledgeable about respiratory illnesses to curb the pandemic (Mohamed et al., 2020). However, lockdowns had negative consequences such as physical inactivity, loneliness, mental stress, depression, anxiety, social dysfunction, learning struggles, and domestic problems (Lima et al., 2020; Lippi et al., 2020). Lockdowns have also negatively affected economies and businesses.

Key lessons learned from COVID-19 relate to our survival, preparedness, and responsibility towards the environment to control future pandemics and strengthen our health infrastructure. Lockdowns have proved to be effective in healing the environment and ecosystem as well as in breaking the chain of infections.

## 16.6 CONCLUSION

The COVID-19 lockdowns resulted in many unforeseeable impacts on the world's economy, environment, and public health. The lockdowns resulted in huge decreases in pollution levels and has provided an opportunity for policy makers, researchers, government, and the public to rethink and make judicious use of natural resources with sustainable development, thereby minimizing global emissions. Increased generation of domestic and bio-medical waste was also seen, which is a matter of concern. Apart from this, the behavioural changes in humans and lesson learned during this pandemic will be helpful for environmental protection and public health policy changes in the future. Moreover, this pandemic has changed people's mindsets and made them consider how they live, with increased affection towards the environment, which will help community engagement and participation to protect the environment.

## REFERENCES

Afelt, A., Frutos, R. and Devaux, C., 2018. Bats, coronaviruses, and deforestation: Toward the emergence of novel infectious diseases? *Frontiers in Microbiology, 9*, p. 702.

Anjum, N.A., 2020. Good in the worst: COVID-19 restrictions and ease in global air pollution. *Preprints* 2020, 2020040069 (doi: 10.20944/preprints202004.0069.v1).

Arora, N.K. and Mishra, J., 2020. COVID-19 and importance of environmental sustainability. *Environmental Sustainability, 3*, pp. 117–119 (https://doi.org/10.1007/s42398-020-00107-z)

Bir, 2020. Qatar: climate change issue demands COVID-19 like measures. www.aa.com.tr/en/environment/qatar-climate-change-issue-demands-covid-19-like-measures-/1809519

Biswal, A., Singh, V., Singh, S., Kesarkar, A.P., Ravindra, K., Sokhi, R.S., Chipperfield, M.P., Dhomse, S.S., Pope, R.J., Singh, T. and Mor, S., 2021. COVID-19 lockdown-induced changes in $NO_2$ levels across India observed by multi-satellite and surface observations. *Atmospheric Chemistry and Physics, 21*(6), pp. 5235–5251.

Blum, B. and Neumärker, B., 2020. Globalization, Environmental Damage and the Corona Pandemic-Lessons from the Crisis for Economic. Environmental and Social Policy. (Available at SSRN: https://ssrn.com/abstract=3613719 or http://dx.doi.org/ 10.2139/ssrn.3613719)

Booth, H., Arias, M., Brittain, S., Challender, D.W., Khanyari, M., Kupier, T., Li, Y., Olmedo, A., Oyanedel, R., Pienkowski, T. and Milner-Gulland, E.J., 2020. Managing wildlife trade for sustainable development outcomes after COVID-19. doi:10.31235/osf.io/2p3xt

Briggs, D., 2003. Environmental pollution and the global burden of disease. *British Medical Bulletin, 68*(1), pp. 1–24.

Cadotte, M., 2020. Early evidence that COVID-19 government policies reduce urban air pollution. https://doi.org/10.31223/osf. io/nhgj3

Carbon Brief, 2020. As China Battles One of the Most Serious Virus Epidemics of the Century, the Impacts on the Country's Energy Demand and Emissions Are Only Beginning to Be Felt. www.eceee.org/all-news/news/analysis-coronavirus-has-temporarily-reduced-chinas-co2-emissions-by-a-quarter/

Chakraborty, I. and Maity, P., 2020. COVID-19 outbreak: Migration, effects on society, global environment and prevention. *Science of the Total Environment, 728*, p. 138882.

Chen, K., Wang, M., Huang, C., Kinney, P.L. and Anastas, P.T., 2020. Air pollution reduction and mortality benefit during the COVID-19 outbreak in China. *The Lancet Planetary Health, 4*(6), pp. e210–e212.

Chinazzi, M., Davis, J.T., Ajelli, M., Gioannini, C., Litvinova, M., Merler, S., Y Piontti, A.P., Mu, K., Rossi, L., Sun, K. and Viboud, C., 2020. The effect of travel restrictions on the spread of the 2019 novel coronavirus (COVID-19) outbreak. *Science, 368*(6489), pp. 395–400.

Coronaviridae Study Group of the International, C.S.G., 2020. The species Severe acute respiratory syndrome-related coronavirus: Classifying 2019-nCoV and naming it SARS-CoV-2. *Nature Microbiology, 5*(4), p. 536.

Dutheil, F., Baker, J.S. and Navel, V., 2020. COVID-19 as a factor influencing air pollution?. *Environmental Pollution (Barking, Essex: 1987), 263*, p. 114466.

ESA. 2020a. www.esa.int/Applications/Observing_the_Earth/ Copernicus/Sentinel-5P/COVID-19_nitrogen_dioxide_over_China

ESA. 2020b. www.esa.int/Applications/Observing_the_Earth/ Copernicus/Sentinel-5P/Coronavirus_lockdown_leading_to _drop_ in_pollution_across_Europe

Gautam, S., 2020. COVID-19: Air pollution remains low as people stay at home. *Air Quality, Atmosphere & Health*, *13*, pp. 853–857.

Helm, D., 2020. The environmental impacts of the coronavirus. *Environmental & Resource Economics*, *76*, pp. 21–38.

Isaifan, R.J., 2020. The dramatic impact of Coronavirus outbreak on air quality: Has it saved as much as it has killed so far?. *Global Journal of Environmental Science and Management*, *6*(3), pp. 275–288.

Kohli, A. 2020. Because of no pollution I can see... hilarious trend takes over Twitter. After pictures of the Dhauladhar mountain range went viral, Twitter users began posting about all the other things they could see from their own homes. www. hindustantimes.com/it-s-viral/because-of-no-pollution-i-can-see-hilarious-trend-takes-over-twitter/story-CykQiSGlY7st GJ1AeU2ksL.html

Lima, N.N.R., de Souza, R.I., Feitosa, P.W.G., de Sousa Moreira, J.L., da Silva, C.G.L. and Neto, M.L.R., 2020. People experiencing homelessness: Their potential exposure to COVID-19. *Psychiatry Research*, p. 112945.

Ling, G.H.T. and Ho, C.M.C., 2020. Effects of the coronavirus (COVID-19) pandemic on social behaviours: From a social dilemma perspective. *Technium Social Sciences Journal*, *7*(1), pp. 312–320.

Lippi, G., Henry, B.M., Bovo, C. and Sanchis-Gomar, F., 2020. Health risks and potential remedies during prolonged lockdowns for coronavirus disease 2019 (COVID-19). *Diagnosis*, *7*(2), pp. 85–90.

Mandal, I. and Pal, S., 2020. COVID-19 pandemic persuaded lockdown effects on environment over stone quarrying and crushing areas. *Science of the Total Environment*, *732*, p. 139281.

Mohamed, A.A.O., Elhassan, E.A.M., Mohamed, A.O., Mohammed, A.A., Mahgoop, M.A., Sharif, M.E., Bashir, M.I., Abdelrahim, R.B., Idriss, W.I. and Malik, E.M., 2021. Knowledge, attitude and practice of the Sudanese people towards COVID-19: An online survey. *BMC Public Health*, *21*(1), pp. 1–7.

NASA, 2020. Airborne Nitrogen Dioxide Plummets over China. https://earthobservatory.nasa.gov/images/146 362/airborne-nitrogen-dioxide-plummets-over-china.

Olivero, J., Fa, J.E., Real, R., Márquez, A.L., Farfán, M.A., Vargas, J.M., Gaveau, D., Salim, M.A., Park, D., Suter, J. and King, S., 2017. Recent loss of closed forests is associated with Ebola virus disease outbreaks. *Scientific Reports*, *7*(1), pp. 1–9.

Ouhsine, O., Ouigmane, A., Layati, E., Aba, B., Isaifan, R. and Berkani, M., 2020. Impact of COVID-19 on the qualitative and quantitative aspect of household solid waste. *Global Journal of Environmental Science and Management*, *6* (Special Issue (Covid-19)), pp. 41–52.

Paital, B., 2020. Nurture to nature via COVID-19, A self-regenerating environmental strategy of environment in global context. *Science of The Total Environment*, p. 139088.

Poudel, B.S., 2020. Ecological solutions to prevent future pandemics like COVID-19. *Banko Janakari*, *30*(1), pp. 1–2.

Rosenbloom, D. and Markard, J., 2020. A COVID-19 recovery for climate. *Science*, 368(6490), pp. 447 Doi: 10.1126/ science.abc4887

Saadat, S., Rawtani, D. and Hussain, C.M., 2020. Environmental perspective of COVID-19. *Science of the Total Environment*, p. 138870.

SAFAR – India (2020) System of Air Quality and Weather Forecasting and Research. http://safar.tropmet. res.in/

Saleh, S., Shepherd, W., Jewell, C., Lam, N.L., Balmes, J., Bates, M.N., Lai, P.S., Ochieng, C.A., Chinouya, M. and Mortimer, K., 2020. Air pollution interventions and respiratory health: A systematic review. *The International Journal of Tuberculosis and Lung Disease*, *24*(2), pp. 150–164.

Schuster, 2020. Coronavirus Lockdown Gives Animals Rare Break from Noise Pollution. www.dw.com/en/ coronavirus-lockdown-gives-animals-rare-break-from-noise-pollution/a-53106214

Sharma, S., Zhang, M., Gao, J., Zhang, H. and Kota, S.H., 2020. Effect of restricted emissions during COVID-19 on air quality in India. *Science of the Total Environment*, *728*, p. 138878.

Shereen, M.A., Khan, S., Kazmi, A., Bashir, N. and Siddique, R., 2020. COVID-19 infection: Origin, transmission, and characteristics of human coronaviruses. *Journal of Advanced Research*, *24*, pp. 91–98.

Singhal, T., 2020. A review of coronavirus disease-2019 (COVID-19). *The Indian Journal of Pediatrics*, *87*(4), pp. 281–286.

Sohrabi, C., Alsafi, Z., O'Neill, N., Khan, M., Kerwan, A., Al-Jabir, A., Iosifidis, C. and Agha, R., 2020. World Health Organization declares global emergency: A review of the 2019 novel coronavirus (COVID-19). *International Journal Of Surgery*, *76*, pp. 71–76.

*The Brussels Time*. 2020. Coronavirus: Nearly a Billion People in Lockdown. Available at: www.brusselsti mes.com/all-news/world-all-news/102043/nearly-a-billion-people-in-lockdownacross-the-world/. Last accessed: March 25, 2020.

Wang, H., Shao, J., Luo, X., Chuai, Z., Xu, S., Geng, M. and Gao, Z., 2020. Wildlife consumption ban is insufficient. *Science*, *367*(6485), pp. 1435–1435.

Watts, 2020. Climate Crisis: In Coronavirus Lockdown, Nature Bounces Back – but for How Long? *The Guardian*, www.theguardian.com/world/2020/apr/09/climate-crisis-amid-coronavirus-lockdown-nature-bounces-back-but-for-how-long

World Health Organization (WHO). 2018. www.who.int/news-room/fact-sheets/detail/ambient-(outdoor)-air-quality-and-health

Wright, R., 2020. The World's Largest Coronavirus Lockdown is Having a Dramatic Impact on Pollution in India. https://edition.cnn.com/2020/03/31/asia/coronavirus-lockdown-impact-pollution-india-intl-hnk/index.html

WHO Report, 2022. World Health Organization, COVID-19 Weekly Epidemiological Update. Edition 117. www.who.int/publications/m/item/weekly-epidemiological-update-on-covid-19---9-november-2022. Last accessed: November 13, 2022.

Yang, N., Liu, P., Li, W. and Zhang, L., 2020. Permanently ban wildlife consumption. *Science*, *367*(6485), pp. 1434–1434.

Zambrano-Monserrate, M.A., Ruano, M.A. and Sanchez-Alcalde, L., 2020. Indirect effects of COVID-19 on the environment. *Science of the Total Environment*, *728*, p. 138813.

# 17 Bio-Medical Waste Management in Post COVID-19 Pandemic

## Need for Effective Regulation in India

*Anis Ahmad*

Department of Law, Babasaheb Bhimrao Ambedkar University, Lucknow, Uttar Pradesh, India

## CONTENTS

## 17.1 INTRODUCTION

In the past few decades, the world has confronted different kinds of environmental challenges such as water pollution, air pollution, conservation of wildlife, protection of birds and animals, deforestation, hazardous waste, conservation of biodiversity and climate change, etc. Among them is the management of hazardous waste including bio-medical waste, which is a serious threat to the ecosystem and survival of human beings on this planet (Sashtri, 2018). Technological advancements and innovative research in the field of medical sciences are no doubt a boon to mankind, but they have also increased the generation of infectious and hazardous bio-medical waste in huge quantities. This poses serious challenges to poor and frail public healthcare systems, particularly in developing countries like India (Jariwala, 1999). COVID-19 has made the problem worse.

In order to address bio-medical waste many international and national regulatory frameworks and strategic plans have been put in place for protecting and improving the environment, and nation states have adopted various measures such as administrative, regulatory, and legal mechanisms to address the above problems (Birnie, 2009). The Government of India has taken several legal and administrative measures for environmental protection and improvement. It was only in the year 1989 in sections 6, 8, and 25 of the *Environment (Protection) Act*, 1986 was *Hazardous Wastes (Management and Handling) Rules,* 1989 included that cover bio-medical waste under hazardous wastes substances. In 1998, the Central Government of India came out with a specific draft rule, *Biomedical Waste (Management & Handing) Rules*, 1998 (Singh, 2016). Thereafter, various amendments took place in these rules from time to time for effective management of bio-medical waste. Further, the Government of India drafted the *Biomedical Waste Management Rules*, 2016. The BMWM rules were further amended in 2018 and 2019 (Kharat, 2016). These rules fill the gaps in the old rules to regulate the disposal of various categories of bio-medical waste. Apart from that, in the recent past the Central Pollution Control Board issued various guidelines for disposal of wastes particularly bio-medical waste during the COVID-19 pandemic in treatment, diagnosis, and quarantine of patients (Guidelines-Covid_1, 2020). The Supreme Court of India and National Green Tribunal also taken addressed the problem of handling, management, and disposal of bio-medical waste. With the above in mind, this chapter examines the existing regulatory and administrative initiatives as well as juristic observations related to bio-medical waste for achieving the sustainable development goals in coming years. Research has been done on the regulation of bio-medical waste, but in the present context of COVID-19 pandemic there is a need to revisit the potentiality of existing law and policy for sustainable management of bio-medical waste for the benefit all.

## 17.2  DANGERS OF BIO-MEDICAL WASTE

The impact of unsustainable management of waste on human health was realized by the World Health Organization (WHO) in 1996. According to the WHO, the global life expectancy is increasing year after year. However, deaths due to infectious disease are increasing. It also reveals that more than 50,000 people die every day from infectious diseases. One of the causes for the increase in infectious diseases is improper waste management. Blood, body fluids, and body secretions, which are constituents of bio-medical waste, harbour viruses, bacteria, and parasites that cause infection. It passes via a number of human contacts, all of whom are potential "recipients" of the infection. Human immunodeficiency virus (HIV) and hepatitis viruses are example infections and diseases documented to have spread through bio-medical waste. Tuberculosis, pneumonia, diarrhea diseases, tetanus, whooping cough, etc., are other common diseases spread due to improper waste management.

In the present context, bio-medical waste management has become a major concern for India taking into account the ongoing COVID-19 pandemic. All bio-medical waste is being disposed of along with municipal solid waste due to inadequate management and lack of awareness at all level of policy decisions. A nationwide survey by the International Clinical Epidemiology Network showed that improper pretreatment of bio-medical waste at source and improper terminal disposal are the major challenges. In the absence of effective and credible bio-medical waste management, there is around 82% of primary, 60% of secondary, and 54% of tertiary care health facilities (www. indiaenvironmentportal.org.in/).

According to the Central Pollution Control Board, the generation of COVID-19 related Bio-medical Waste in States/UTs as on 05/01/2021 about 184 out of 198 Common Bio-medical Waste Treatment Facilities (CBWTFs) are using the Tracking app. Around 8,000 generators of COVID-19 waste have registered and used the Tracking App. As per the information submitted by the State Pollution Control Boards/Control Committees as well as daily data received from COVID-19 bio-medical waste tracking app, the average quantity of COVID-19 related bio-medical waste generation was about 146 TPD for the month of December 2020. About 198 of CBWTFs (almost all facilities) are involved in treatment & disposal of COVID-19 bio-medical waste across the country. The present status of COVID-19 bio-medical waste generation and treatment facilities involved in collection, treatment and disposal for the month of December 2020 in 35 States and Union Territories, the bio-Medical waste generation during COVID-19 BMW is total 4527.55 tonnes in Total no. 198 CBWTFs are engaged. In states Maharashtra is the top and then Kerala and among the Union Territories Delhi is top in rank and then Chandigarh in generation of bio-medical waste (https://cpcb.nic.in/uploads/Projects/Bio-Medical-Waste/COVID19_Waste_Management_status_December2020.pdf).

Further, as per the report of the Associated Chambers of Commerce & Industry of India, which was released in 2018, India is likely to generate about 775.5 tons of medical waste per day by 2022. The waste management market in India is expected to reach USD 13.62 billion by 2025 (*Economic Times*, 23th March 2018). Moreover, it is surprising that per the Annual Report 2020–2021 by the Ministry of Environment, Forest and Climate Change the Government of India published in October 2021, there is no data related to bio-medical waste generation and its regulation rather it is only directed to stakeholders to prepared technical guidelines for the implementation of *Biomedical Waste Management Rules, 2016* (http://moef.gov.in/en/annual-report-2020-21/).

## 17.3 BIO-MEDICAL WASTE MANAGEMENT: CONCEPTUAL FRAMEWORK

The whole concept of waste management involves appropriate mechanisms to reduce the amount of waste generated, manage the unavoidable waste to avoid unpremeditated pollution, and establish an effective e recycling system (Ansari, 2012).

According to the Basel Convention, which defines wastes that the "Wastes are such substance, which is intended and essential to be disposed of by the provisions of national legal instruments." It can be categorized into different types based on their source, and encompasses innumerable wastes like household waste, industrial waste, bio-medical waste, and hospital waste (Article 2, Basel Convention).

The WHO defines bio-medical waste as waste generated by healthcare facilities such as from physician's offices, hospitals, dental practices, laboratories, medical research facilities, and veterinary clinics.

In India, the *Biomedical Waste (Management & Handing) Rules,* 1998 amended in 2016, defines "bio-medical waste" as any waste that is generated during the diagnosis, treatment, or immunization of human beings or animals or research activities pertaining thereto or in the production or testing of biological or in health camps, including the categories mentioned in Schedule I appended to these rules (Rule 3(f)2016).

## 17.4 LEGISLATIVE DEVELOPMENTS

### 17.4.1 CONSTITUTIONAL PRECEPTS

The Constitutional provisions provide the bed-rock for the framing of the environmental legislation in the country (Bakshi, 2006). The Constitution of India was adopted in 1950, but it did not have any specific provision with regard to protection of the environment (Venkat, 2011), since environment per se was not enumerated in either of the lists in the Seventh Schedule of the Constitution (Nomani, 2004). It was only in the Forty-second Amendment Act, 1976 that two articles were inserted into the Constitution. The mandatory language employed in Article 48-A, while constitutionally obligated the state to endeavor protect and improve the country's environment and to safeguard its forests and wildlife, similar mandatory language was employed in Article

51-A (g) that obligated every citizens of India has the duty to protect and improve county's natural environment including its forests, lakes, rivers, and wildlife and to have compassion for living creatures (Shyam and Rosencranz, 2001). To ensure that fundamental rights are properly protected, the Constitution conferred on the Supreme Court and High Courts of India the power to determine the most effective remedies whenever such rights are violated (Reddy, 2011). Under the provisions of Articles 32 and 226 the Courts have liberally construed and expanded the scope of fundamental rights to a wholesome environment as part of the right to life guaranteed in Article 21 of the Constitution that encompass a group rights known as third generation rights (Sharifuddin, 2011).

### 17.4.2 Bio-Medical Waste Management Rules in India

In 1998, the Government of India under sections 6, 8 and 25 of the *Environment (Protection) Act, 1986* passed the *Bio-Medical Waste (Management and Handling) Rules, 1998* with an aim to improve the collection, segregation, processing, treatment, and disposal of bio-medical wastes in an environmentally sound way, thereby reducing bio-medical waste generation and impact on environment (Preamble of the Rule, 2016). It consists of thirteen rules and two schedules. Various amendments to these rules have been made. These amendments consist of eighteen rules, four schedules, and five forms. These rules apply to all people who generate, collect, receive, store, transport, treat, dispose, or handle bio-medical waste in any form (Rule 2(1) 2016). But are not applicable to radioactive wastes, hazardous chemicals, and solid waste, lead-acid batteries, hazardous waste, waste covered under the e-waste, hazardous microorganisms, and genetically engineered microorganisms. The parent rule of 1998 is silent about the Common Bio-medical Waste Treatment Facility (CBMWTF), which is defined by the 2016 amended rule as a person who owns or controls a CBMWTF for the collection, reception, storage, transport, treatment, disposal, or any other form of handling of bio-medical waste (Rule 2 (n) 2016). The duties of the operators of a common bio-medical waste treatment and disposal facility as occupier of the treatment plant (Rule 5, 2016). The rules define the occupier (Rule 2 (m) 2016) as a person having administrative control over the institution and the premises generating bio-medical waste. It shall be the duty (Rule 4,2016) of every occupier that to take all necessary steps to ensure that bio-medical waste is handled without any adverse effect to human health and the environment, segregated bio-medical waste in colored bags, phase out the use of chlorinated plastic bags, not to give treated bio-medical waste with municipal solid waste, provide training to all its health care workers, establish a bar-code system for bags or containers and maintain all record for the operation of incineration. Further, it is provided that every occupier or operator of a CBMWTF shall submit an annual report to the prescribed authority in Form-IV, on or before the 30th June of every year (Rule 13,2016). For the proper execution of such rules, it further provides the prescribed authority for the State Pollution Control Board shall be the prescribed authority for the state and Pollution Control Committees in respect of Union territories (Rule 9, 2016). It shall comply with the responsibilities as stipulated in Schedule III of the rule. Further is to provide that every State Government or Union territory Administration shall Constitute an Advisory Committee for the respective State or Union territory under the chairmanship of the respective health secretary to oversee the implementation of the rules (Rule 11,2016. Every occupier or operator handling bio-medical waste shall make an application in Form II to the prescribed authority for grant of authorization and the prescribed authority shall grant the provisional authorization in Form III (Rule 10,2016). It is binding that the application must be disposed of by prescribed authority within 90 days from the date of receipt of a duly completed application. Any person aggrieved by an order made by the prescribed authority under these rules may prefer an appeal (Rule 16, 2016) in Form V to the Secretary (Environment) of the State Government or Union territory administration within a period of thirty days from the date on which the order is communicated to him. For the proper management of bio-medical waste, the rule provides the treatment and disposal (Rule 7, 2016) of it under the first schedule in compliance with the standards provided in the second schedule by the health care facilities and CBMWTF. The bio-medical waste shall be segregated (Rule 8, 2016) into containers or bags at the point of generation. The Ministry of Environment, Forest and Climate Change shall review the implementation of the rules in the country once in a year through the State Health Secretaries and Chairmen or Member Secretary of State Pollution Control Boards and Central Pollution Control Board and the Ministry may also invite experts in the field of bio-medical waste management (Rule 12, 2016). Every authorized person shall maintain records related to the generation, collection, reception, storage, transportation, treatment, disposal, or any other form of handling of bio-medical waste, for five years, following these rules and guidelines issued by the Central Government or the Central Pollution Control Board or the prescribed authority(Rule 14, 2016). Further, it also provides that In case of any major accident at any institution or facility or any other Site while handling bio-medical waste, the authorized person shall intimate immediately to the prescribed authority about such accident and forward a report within 24 hours in writing regarding the remedial steps taken (Rule 15, 2016).

The bio-medical waste management rules were further amended in March 2018 with the following following duties of occupiers/operators prescribed under BMWM Rules 2015 are amended:

   i. Healthcare Facilities shall phase out of chlorinated plastic bags (excluding blood bags) and gloves by 27th March, 2019.
  ii. Establish a bar-code system for bags or containers containing biomedical waste to be sent out of premises for the further treatment & disposal of biomedical waste by 27.03.2019.
 iii. All Healthcare Facilities (any no. of bed) shall make available the annual report on its website by 15.03.2020 (CPCBthe Annual Report 2017–18).

### 17.4.3 INSTITUTIONAL STRATEGIES DURING COVID-19 PANDEMIC

The *Biomedical Waste Management Rules*, 2016 mandates the Central Pollution Control Board (CPCB) to publish guidelines to facilitate implementation by stakeholders including State Pollution Control Boards/PCCs. Hence, the CPCB has prepared a number of guidelines for "Management of Healthcare Waste by the Healthcare Facilities as per Biomedical Waste Management Rules, 2016."

In order to deal with the COVID-19 Pandemic, state and central governments have initiated various steps, which include setting up of quarantine centers/isolation wards, sample collection centers, and laboratories (Guidelines - 17th July, 2020 Guidelines for Handling, Treatment and Disposal of Waste Generated during Treatment/Diagnosis/Quarantine of COVID-19 Patients, available at: https://cpcb.nic.in/uploads/Projects/Bio-Medical-Waste/BMW-GUIDELINES-COVID_1.pdf).

Specific guidelines for management of waste generated during diagnostics and treatment of COVID-19 suspected/confirmed patients are required to be followed by all stakeholders including isolation wards, quarantine centers/isolation wards, sample collection centers and laboratories, Urban Local Bodies (ULBs), and common bio-medical waste treatment and disposal facilities, in addition to existing practices under *Bio-Medical Waste Management Rule, 2016*. These guidelines are based on current knowledge of COVID-19 and existing practices in management of infectious waste generated in hospitals treating viral and other contagious diseases. These guidelines were based on the guidelines issued by the WHO, MoH&FW, ICMR, CDC, and other concerned agencies.

The following salient features of the guidelines are given below:

### 17.4.3.1   COVID-19 Isolation Wards: (Isolation Wards are Those Where COVID-19 Positive Patients are Being Kept for Treatment/Diagnosis)

• Keep separate color coded bin/bags/containers in wards and maintain proper segregation of waste as per BMWM Rules, 2016 as amended and CPCB guidelines for implementation of BMW management Rules.

- As precaution double layered bags (using 2 bags) should be used for collection of waste from COVID-19 isolation wards so as to ensure adequate strength and no-leaks.
- Collect and store biomedical waste separately prior to handing over the same CBWTF. Use a dedicated collection bin labeled as "COVID-19" to store COVID-19 waste and keep separately in temporary storage room prior to handing over to authorized staff of CBWTF. Biomedical waste collected in such Isolation wards can also be lifted directly from ward into CBWTF collection van.
- In addition to mandatory labeling, bags/containers used for collecting biomedical waste from COVID-19 wards, should be labeled as "COVID-19 Waste." This marking would enable CBWTFs to identify the waste easily for priority treatment and disposal immediately upon the receipt.
- Maintain separate record of waste generated from COVID-19 isolation wards.
- Use dedicated trolleys and collection bins in COVID-19 isolation wards.
- The (inner and outer) surface of containers/bins/trolleys used for storage of COVID-19 waste should be disinfected with 1% sodium hypochlorite solution daily.
- Report opening or operation of COVID-19 ward and COVID ICU ward to SPCBs and respective CBWTF located in the area.
- Depute dedicated sanitation workers separately for biomedical waste so that waste can be collected and transferred timely to temporary waste storage area.
- Feces form COVID-19 confirmed patient, who is unable to use toilets and excreta is collected in diaper, must be treated as biomedical waste and should be placed in yellow bag/container. However, if a bedpan is used, then faces to be washed into toilet and cleaned with a neutral detergent and water, disinfected with a 0.5% chorine solution, then rinsed with clean water.
- Collect use PPEs such as goggles, face-shield, splash proof apron, Plastic Coverall, Hazmet suit, nitrile gloves into red bag.
- Collect used masks (including triple layer mask, N95 mask, etc.), head cover/cap, shoe-cover, disposable linen gown, non-plastic or semi-plastic coverall in yellow bags.
- Used masks, tissues, and toiletries, of COVID-19 patient shall become biomedical waste and shall be segregated in yellow bag.

### 17.4.3.2 Sample Collection Centers and Laboratories for COVID-19 Suspected Patients

- Report opening or operation of COVID-19 sample collection centers and laboratories to concerned SPCB. Guidelines given at section (a) for isolation wards should be applied suitably in case of test centers and laboratories. Pre-treat viral transport media, plastic vials, vacutainers, eppendorf tubes, plastic cryovials, pipette tips as per BMWM Rules, 2016 and collect in red bags.

### 17.4.3.3 Responsibilities of Persons Operating Quarantine Camps/Homes or Home-Care Facilities

Less quantity of bio-medical waste is expected from quarantine camps/quarantine homes/home-care facilities. However, the people responsible for operating these facilities need to follow these steps to ensure safe handling and disposal of waste:

- General solid waste (household waste) generated from quarantine centers or camps should be collected in bags, securely tied and handed-over to municipal solid waste collector identified by Urban Local Bodies for final disposal.
- General solid waste should comprise of waste generated from kitchen, packaging material, waste food material, waste papers, waste plastics, floor cleaning dust, etc. including left-over food, disposable utensils, water bottles, tetra packs, used by suspected quarantined persons and COVID-19 patient at homecare or home quarantine.

- Only the used masks, gloves and tissues or swabs contaminated with blood/body fluids of COVID-19 patients, including used syringes, medicines, etc., if any generated should be treated as biomedical waste.
- Biomedical waste if any generated from quarantine centers/camps should be collected separately in yellow colored bags (suitable for biomedical waste collection) provided by ULBs. These bags can be placed in separate and dedicated dust-bins of appropriate size.
- Persons operating quarantine camps/centers should call the CBWTF operator to collect biomedical waste as and when it gets generated. Contact details of CBWTFs would be available with local authorities.
- Persons taking care of quarantine home/home-care should deposit biomedical waste if any generated from suspected or recovered COVID-19 patients.

### 17.4.3.4 Duties of Common Biomedical Waste Treatment Facility (CBWTF)
- Report to SPCBs/PCCs about receiving of waste from COVID-19 isolation wards/ Quarantine;
- Camps/quarantined homes/COVID-19 testing centers;
- Operator of CBWTF shall ensure regular sanitization of workers involved in handling and collection of biomedical waste;
- Workers shall be provided with adequate PPEs including three layer masks, splash proof aprons/gowns, nitrile gloves, gum boots and safety goggles;
- Use dedicated vehicle to collect COVID-19 ward waste. It is not necessary to place separate label on such vehicles;
- Vehicle should be sanitized with sodium hypochlorite or any appropriate chemical disinfectant after every trip.
- COVID-19 waste should be disposed-off immediately upon receipt at facility;
- In case it is required to treat and dispose more quantity of biomedical waste generated from
- COVID-19 treatment, CBWTF may operate their facilities for extra hours, by giving information to SPCBs/PCCs;
- Operator of CBWTF shall maintain separate record for collection, treatment, and disposal of COVID-19 waste;
- Do not allow any worker showing symptoms of illness to work at the facility. May provide an adequate leave to such workers and by protecting their salary;
- CBWTF operator shall register on "COVID19BWM" Tracking App developed by CPCB and also ensure registration of Waste Handler (with vehicle) for entering the data of COVID-19 biomedical waste received and disposed;
- Provide training to Waste handlers on infection prevention measures, hand hygiene, respiratory etiquettes, social distancing, use of PPE, etc. via videos and demonstrations etc. translated in local language. Sanitation workers more than 50-year of age should be posted for management of non-COVID waste.

### 17.4.3.5 Duties of SPCBs/PCCs
- Shall maintain records of COVID-19 treatment wards/quarantine centers/quarantines homes in respective states.
- Ensure proper collection and disposal of biomedical waste as per BMW Rules, 2016 and SOPS given in this guidance document;
- Allow CBWTFs to operate for extra hours as per requirement;
- May not insist on authorisation of quarantine camps as such facilities does not qualify as health facilities. However, may allow CBWTFs to collect biomedical waste as and when required;

- In case of States not having CBWTFs as well as rural or remote areas, not having access to CBWTFs, the existing captive facilities of any hospital may be identified for disposal of COVID-19 waste as per provisions under BMWM Rules, 2016 and these guidelines. This may include permitting use of deep burial pits for disposal of yellow category waste as per standards prescribed in Schedule II of Bio-medical Waste Management Rules, 2016.
- Coordinate with CBWTFs and ULBs in establishing adequate facilities for collection and disposal of COVID-19 waste.
- In case of generation of large volume of yellow color coded (incinerable) COVID-19 waste, permit HW incinerators at existing TSDFs to incinerate the same by ensuring separate arrangement for handling and waste feeding.
- During COVID-19 pandemic, SPCBs/PCCs may direct the ULBs to collect dry general solid waste in bags from quarantine Centers/Quarantine homes/Homecare units, and sprayed with disinfectant solution, for disposal in waste to energy plants/industrial incinerators/landfills, as per existing practice or availability in the state.
- Every SPCB/PCC shall use "COVID19BWM" web-portal developed by CPCB to track and verify COVID-19 biomedical waste and to submit daily data to CPCB through said portal.

### 17.4.3.6   Duties of Urban Local Bodies

Urban local bodies are responsible for ensuring safe collection and disposal of bio-medical waste, if any, generated from quarantine camps/quarantine homes/homecare facilities for COVID-19 patients.

- Information on each Quarantine Camps/Quarantine Homes/Home-Care should be available with local administration and provide updated list to SPCBs from time to time;
- In case of quarantine camps, ensure that biomedical waste is collected directly by CBWTFs identified by ULB. Waste from quarantine camps to be lifted by CBWTFs on call basis as and when the biomedical waste gets generated. Provide contact details of CBWTF operator at Quarantine Camps;
- Provide necessary support, security including authorisation to staff of CBWTFs;
- ULB shall engage CBWTF operator for ultimate disposal of biomedical waste collected from quarantine home/home care or waste deposition centers or from doorsteps as may be required depending on local situation; ULB shall make agreement with CBWTF in this regard;
- Ensure that general solid waste and biomedical waste generated from quarantine camps/quarantine homes/Homecare is not mixed. The biomedical waste and general solid waste should be collected separately. Inform the persons responsible for operating isolation wards, quarantine centers and residents of homecare units to collect solid waste and biomedical waste in separate bags securely tied prior to hand over to authorized waste collectors of ULBs. ULBs should ensure that left-over food and general solid waste is not collected in yellow bags;
- Create awareness among citizens regarding segregation of municipal solid waste and biomedical waste (as part of Domestic Hazardous Waste) generated from homes/quarantine homes/home care facilities;
- Services of Common Biomedical Waste Treatment & Disposal Facilities (CBWTFs) and staff associated with CBWTFs for collection, transportation, treatment, and disposal of biomedical waste generated from hospitals including COVID-19 isolation wards, Quarantine Camps, etc., may be considered an essential service as part of health infrastructure;
- Facilitate smooth operations of CBWTFs;
- Local agencies/ULBs may take additional measures considering prevailing ground situations and feasibility, however while implementing such measures requirements outlined in these guidelines should be complied;

- ULB shall designate a Nodal person who will be responsible for waste management for specific quarantine center or an area and for maintenance of its record;
- Designated nodal person shall download and use biomedical waste Tracking App "COVID19BWM" developed by CPCB to feed daily data on quantity of biomedical waste collected from home-cares or home quarantines;
- In case ULBs are unable to manage solid waste with their existing staff, professional solid waste management agencies may be engaged/authorized during COVID-19 situation for timely collection of solid waste and biomedical wastes separately from quarantine centers, home-care, COVID-19 isolation wards, and quarantine homes.

### 17.4.3.7 Management of Wastewater from HCFs/Isolation Wards

As per the information available from the US CDC, the risk of transmission of the virus that causes COVID-19 through sewerage systems is thought to be low. Transmission to operators may be possible during treatment of sewage treatment plants; however, there is no evidence to date that this has occurred. Therefore, the following guidance is recommended for HCFs and the operator of STPs:

- Responsible agencies are Healthcare Facilities/Isolation Wards/operators of terminal sewage treatment plants (PHED/Jal Board/etc.).
- HCFs and the agencies operating Sewage Treatment Plants should continue to ensure disinfection of treated wastewater as per prevailing practices to inactivate corona viruses.
- Operators of ETPs/STPs attached with discharge from Healthcare Facilities and isolation wards should adopt standard operational practices, practice basic hygiene precautions, and wear personal protective equipment (PPE) prescribed for operation of STPs. PPEs should include Goggles, face mask, liquid repellant coveralls, waterproof gloves and rRubber boots.
- During the period of COVID-19 pandemic, utilization of treated wastewater in utilities within HCFs may be avoided.

### 17.4.3.8 Disposal of Used PPEs

- Waste masks and gloves in general households should be kept in paper bag for a minimum of 72 hours prior to disposal of the same as dry general solid waste after cutting the same to prevent reuse.
- Discarded PPEs from general public at commercial establishments, shopping malls, institutions, offices, etc. should be stored in separate bin for 3 days, thereafter disposed of as dry general solid waste after cutting/shredding.
- At Material Recovery Facilities (MRFs), discarded PPEs containing plastic should be shredded and sent to SPCB authorised plastic waste recyclers, or may be converted into refuse derived fuel (RDF) for co-processing or energy recovery (in Waste to Energy Plants) or for road making. Shredded PPEs may be disposed at landfill only in case the requisite infrastructure as required under SWM Rules is not available in the State.
- PPEs doffed by healthcare workers accompanying diseased body of COVID-19 patient to crematorium/graveyards should be treated as biomedical waste and disposed as per provisions under SWM Rules, 2016 and BMW Management Rules, 2016. Crematoriums/graveyards may opt for disposal of such PPEs is given below;
- Should be collected in separate bin with yellow-bag and handed over to authorized waste picker engaged by of ULBs for disposal through CBWTFs. Or
- Dispose as domestic hazardous waste (biomedical waste) and may be deposited at designated deposition centers identified by ULBs for pick-up by CBWTFs.

## 17.5  JUDICIAL RESPONSE AND NATIONAL GREEN TRIBUNAL DIRECTIONS

India has enacted various laws at regular intervals to deal with the protection and improvement of the environment and created different rules for better environmental governance from 1974 onward. At the same time the Indian judiciary has played a pivotal role in interpreting the laws in such a manner that not only helped in protecting the environment but also in promoting the concept of sustainable development (Kumar, 2014).

The Supreme Court of India for the first time in 1994 in *Dr. B.L. Wadhera v. Union of India* ((1996) 2 SCC 594) attracted the attention of the apex court when a writ petition was filed under 32 of the Constitution of India. The petitioner sought necessary directions to the Municipal Corporation, Delhi (MCD) and the New Delhi Municipal Council (NDMC) to perform their statutory duties, in particular the collection, removal, and disposal of garbage and other wastes. On the other hand, MCD and NDMC argued non-availability of funds, inadequacy or inefficiency of staff, and insufficiency of machinery for the non-performance of their duties. The Court considered all problems of management of solid waste in cities and in an effort to see that the Capital of India is not branded as being one of the most polluted cities in the world, issued valuable directions for the disposal of their garbage and hospital waste.

In *Almitra H. Patel v. Union of India.* ((2000) 2 SCC 679) the Court after expressing unhappiness over the high levels of pollution in Delhi and the lack of accountability at all levels of the municipal authority concerned and over the inaction of respondent authorities in connection with the directions issued in Dr. B.L. Wadhera's case issued additional directions.

In *Environment Monitoring Forum and Anr. v. Union of India (UOI) and Ors* (MANU/KE/0894/2003) the Court held that it is the duty of the institutions generating bio-medical waste to take all steps to ensure that such waste is handled without any adverse effect on human health and environment.

In *Maitree Sansad v. State of Orissa and Ors* (15th Nov, 2006) the Court observed that the improper practices such as dumping of bio-medical waste in municipal dustbins, open spaces, water bodies, etc., leads to spread of diseases. Emissions from incinerators and open burning also lead to exposure to harmful gases, which can cause cancer and respiratory diseases. Exposure to radioactive waste can also cause serious health hazards. An often-ignored area is an increase in in-home healthcare activities. An increase in the number of diabetics who inject themselves with insulin, home nurses taking care of terminally ill patients, etc., all generate bio-medical waste, which can cause health hazards. The Court further directed all the hospitals in City of Cuttak to strictly comply with the provisions of the Rules, 1998 and if not, the steps in this regard to be taken by the State Pollution Control Board.

Apart from that the role of the National Green Tribunal over the past few years suggests that it has been progressive in its approach towards environmental protection and sustainable development. The National Green Tribunal has been given the power to regulate the procedure by itself. It does not follow the principles of civil procedure code but instead follows the principles of natural justice.

In *Krishan Lal Gera v. State of Haryana and Ors* (*MANU/GT/0140/2015*) the Principal Bench of National Green Tribunal New Delhi Physical directed that injuries may occur to the hospital personnel as well as waste handlers outside the hospital due to improper handling of various bio-medical wastes. Chemical injuries can occur due to hazardous, toxic, corrosive, flammable, and reactive and genotoxic wastes that are likely to cause chemical burns on accidental exposure, or toxicity to cell cytotoxic materials.

Again, the National Green Tribunal Principal Bench, New Delhi in *Haat Supreme Wastech Pvt. Ltd. & Ors. v. State of Haryana & Ors* (MANU/GT/0089/2015). Justice Swatanter Kumar held that "Units are carrying on the activity of handling bio-medical waste treatment plants shall be required to obtain environmental clearance in as per provision of law."

In *Mahesh Dubey v. Chattisgarh Environment Conservation Board and Ors* (MANU/GT/0140/2016) since the State of Chhattisgarh consists of 27 districts and has only four CBWTFs though there should one such plant every 150 km. The respondents are not even aware of how much bio-medical waste is being generated by healthcare facilities being run in the state.

In the *Shalesh Singh case* (16th Nov, 2017) a U.P.-based journalist seeking directions for closure of all hospitals, medical facilities, and Bio-medical waste disposal plants not complying with the waste management rules. It has alleged that rag pickers were allowed unauthorized transportation of waste and they disposed of it in a unscientific manner. The National Green Tribunal issued notice to Uttar Pradesh, Uttarakhand, Haryana, and Punjab governments over improper collection, segregation, and disposal of bio-medical waste. The Tribunal directed that indiscriminate disposal of bio-medical waste and exposure to such waste poses serious threat to the environment and human health that requires specific treatment and management prior to its final disposal. In non-compliance with the above orders the National Green Tribunal imposes fines of Rupees 15.48 lacs on each hospital for not following the *Bio-Medical Waste Management Rules*, 2016 (17th July, 2019).

In *Amrish Gupta, President, Dushit Paryavaran v. State of Uttar Pradesh*, the National Green Tribunal Principle Bench, New Delhi passed order that a joint report was sought from district Magistrate, Barabanki, CPCB and Up State PCB with reference to the allegation of illegal activities in disposal of infected bio-allegation of illegal activities of untreated hazardous effluents and causing air pollution by M/s Synergy Waste Management Pvt. Ltd., Barabanki, Lucknow, Uttar Pradesh (14th Feb, 2020).

In the *Re:Suomoto Scientific Disposal of Bio-Medical Waste Arising Out of COVID-19 Treatment-Compliance of BMW Rules, 2016* case the issue for consideration was remedial action to address the gaps in compliance of the *Bio-Medical Waste Rules*, 2016, as applicable to the disposal of bio-medical waste arising out of handling of COVID-19 disease, so as to ensure protection of environment and public health.

The National Green Tribunal directed the Chief Secretary of States/UTs to coordinate the activities of the state's concerned departments such as Urban Development, Health, Irrigation & Public Health and to closely monitor the scientific storage, transport, handling, management, and disposal of COVID-19 waste. At the national level, a high-level task team of Ministry of MoEF & CC, Health UD, Jal Shakti, Defence, and CPCB was constituted to supervise the handling and scientific disposal of COVID-19 waste in accordance with the guidelines. State Departments of Environment and PCBs/PCCs ensure compliance of *Biomedical Waste Management Rules*, 2016 and report to CPCB and CPCB as directed to take further steps and provide a consolidated report to the Tribunal of the steps taken and the ground status (23th April, 2020, available at: https://greentribunal.gov.in/case Details/DELHI/0701102004122020?page=order)

## 17.6 CONCLUSION

The management of bio-medical waste in India is a challenge, but it is clear from this chapter that existing regulatory and institutional mechanisms have the potential to combat this environmental issue, which poses a serious threat to human health and environmental sustainability. The ongoing COVID-19 pandemic had made the issue more prevalent as and enforcement agencies such as the Central Pollution Control Board and State Pollution Control Boards have had to ensure bio-medical waste is managed properly to contain the spread of the virus. The National Green Tribunal (NGT) has also actively been involved in responding to the challenge. Analysis of cases has shown that the Supreme Court and NGT are also ensuring compliance and credibility in addressing this global crisis.

## REFERENCES

Abdul Haseeb Ansari, Praveen Jamal and Umar A.Oseni, 2012, "Sustainable Development: Islamic Dimension with Special Reference to Conservation of Environment." *Advances in Natural and Applied Sciences* 6(5), 607–619.

Aruna Venkat, 2011 "Environmental Law and Policy." 51. New Delhi, India: PHI Learning Pvt. Ltd.

Birnie Patricia, Alan Boyle, Catherine Redgwell, et. al. 2009, "International Law and the Environment." 50 London: Oxford University Press.

Central Pollution Control Board, Guidelines – 19th April, 2020. B-31011/BMW (94)/2020/WM-I.Visit. www. cpcbenvis.nic. in/pdf/BMW-GUIDELINES-COVID_1.pdf. (Accessed June 12, 2020).

Central Pollution Control Board, Guidelines – Visit https://cpcb.nic.in/uploads/Projects/Bio-Medical-Waste/ BMW-GUIDELINES-COVID1.pdf. (Accessed April 04, 2021).

Dal Singh Kharat, 2016. "Bio-Medical Waste Management Rules, 2016: A Review." *International Journal of Advanced Research and Development*, 48–51.

Divan Shyam, Armin Rosencranz, 2001. "Environmental Law and Policy in India: Cases, Materials and Statutes." 49 New Delhi, India: Oxford University Press.

Gazette of India, Extraordinary, Part II, Section 3, Sub-section (i) dated 28 March 2016, Bio-Medical Waste Management Rules, 2016, Ministry of Environment and Forests, Government of India, New Delhi, India. 2016.

Gurdip Singh, 2016. "Environmental Law." 232 Lucknow, India: Eastern Book Company

C.M. Jariwala, 1999. The Bio-Medical Waste: Direction of Law and Justice. *Journal of Indian Law Institute* XLI 368.

Kumar Swatanter, 2014. "Access to Environmental Justice in India and Indian Constitution." *NGT International Journal of Environment*, 1.

Mohd. Sharifuddin, 2011. "Human Rights to Environment in India." 125 New Delhi, India: Mittal Publications.

O. Chinnappa Reddy, 2011. "The Court and the Constitution of India: Summits and Shallows."246 New Delhi, India: Oxford University Press.

P.M. Bakshi, 2006. "The Constitution of India." 90–92. New Delhi, India: Universal Law Publishing Co.

Ramaswamy Aarthy, Sambit Dash, 2019. "Neglect of Household Biomedical Waste." *Economic & Political Weekly* LIX 49: 19.

S.C. Shastri, 2018. "Environmental Law" 217 Lucknow, India: Eastern Book Company.

Universal's Environmental Laws, New Delhi, India: Universal Law Publishing Co. Pvt. Ltd.

Zafar Mahfooz Nomani, 2000. "Human Right to Environment in India: From Legality to Reality." Delhi Law Review, XXII, 78–98.

Zafar Mahfooz Nomani, 2004. "Natural Resources Law and Policy." 18 New Delhi, India: Uppal Publishing House.

# 18 COVID-19 Pandemic

## *Social Response through Changing Lifestyles*

*Vibhash C. Jha*
Visva-Bharati University, West Bengal, India

## CONTENTS

## 18.1 INTRODUCTION

History shows our world has been facing pandemics every 100 years. However, with the advancement of civilizations, the severity of threats to humankind has increased proportionately. COVID-19 is proof of that. The World Health Organization (WHO) on March 11, 2020 characterized COVID-19 as a pandemic (pandemonium + epidemic) (Ghebreyesus 2020). In a pandemonium one is faced with a situation that is unprecedented, not experienced before and hence there is no preparedness.

Governments have set up taskforces, allocated massive amounts of funds, and set up research teams to identify, isolate, and come up with preventive measures to protect the citizens of the world.

In epidemics like the Bubonic plague about 2000 years back and the Spanish Flu in 1920 (about a century back) the common and standard protocol was always to isolate and quarantine the infected until cured. If we refer to our *Shastras* (ancient scripts), even the almighty are not spared the ordeal of quarantine with Lord *Jagannatha*, *Subhadra*, and *Balbhadra* catching the flu every *snanpurnima* during the monsoons, being quarantined for 15 days, then recovering and appearing before the devotees on *RathYatra* to visit their Aunt *Gundicha* for 10 days. It should be noted here that during this 15 days of quarantine and recovery the deities are fed fruits and fruit juices only. This is no different from our modern way of looking after the sick by providing easily digestible and wholesome

DOI: 10.1201/9781003358916-18

food to them, to aid recovery. So the concept of quarantine and socially as well as physically distancing ourselves from the sick has been in practice since ancient times. What is required in present times is the need to understand its importance and at the same time understand the society's response to COVID-19 to such concepts as quarantine, lockdown, physical distancing, and reducing social interactions, which will be a crucial method of tackling and reducing the impact of such an epidemics.

Gurudev Rabindranath Tagore has referenced the plague, cholera, and flu pandemics in several of his fiction writings. For example, in his novel *Chaturanga*, the social impact of quarantine during plague pandemic in Calcutta has been depicted and the lead character Jagmohan asks, "Should the sick be treated like criminals?" In another of his novel, Gora, the character Harimohini recounts her personal tragedy of losing her son and husband to cholera within just four days: "How could I enjoy such an excess of good fortune without paying a price?" Tagore had also got actively involved in social service during outbreaks of epidemics in Calcutta, alongside Sister Nivedita in organizing relief and medical aid for the victims of this dread pandemic (Kripalani 2008; Pal 2016). Let's also recall that the founder of Visva-Bharati even donned the physician's role by administering an ayurvedic potion, "Panchatikta Panchan," to his students in Santiniketan to thwart the hazard of the dreaded flu in 1918. His prescribed concoction was made of Neem leaves, Gulancha, Teuri, Thankuni, and Nishinda, and it helped in boosting immunity. The poet wrote about Panchatikta Panchan to the renowned scientist and close friend Jagadish Chandra Bose in 1919. He stated,

> none of the boys (students) has got the influenza. I believe it is so because I have been regularly giving them the Panchatikta Panchan. Many of the boys had suffered from the disease during their stay at home in the holidays. Some of them are from the hotspots of the disease and have survived death. I was afraid that they would inadvertently spread the disease here on their return. But nothing of that sort has happened. Even cases of regular fever are fewer this year. Around two hundred people live here, yet the local hospital has almost no patient. This is rather unusual. I am inclined to believe that this is definitely the miracle of the Panchan.

Tagore's relatives struggled with pneumonia and flu-like symptoms and similar health hazards of the time and he lost his 16 years old youngest son Samindranath, due to cholera epidemic in November 1907, who was visiting Monghyr (Bihar, India).

The COVID-19 pandemic has been particularly damaging because resources were limited for tracing, tracking, travelling, treatment, prevention, mitigation, and preparedness. Relief and rehabilitation aids are therefore an important area for solidarity and support through social response. Covid has changed the lifestyles of people around the world and thus our social responses.

## 18.2   COVID-19 PANDEMIC

The COVID-19 pandemic means that many of us are staying at home and doing less in terms of social interactions and exercise. This can have a negative effect on physical and mental health. Lack of proper information, miscommunication, rumors, and false reporting by media regarding policies, vaccination methods, and availability of vaccines have also contributed negatively to people's well-being. The UN's Framework for the Immediate Socio-Economic Response to the COVID-19 Crisis notes that "The COVID-19 pandemic is far more than a health crisis." The impact of Covid-19 on sociopolitical and economic systems and lifestyles is global, and while the impact of the pandemic will vary from country to country, it will most likely increase poverty and inequalities at a global scale. Sustainable Development Goals (SDGs) (UNDP n.d.) are thus even more important today. Assessing the impacts of the COVID-19 crisis on societies, economies, and vulnerable groups is fundamental to inform and tailor the responses of governments and partners to recover from the crisis and ensure that no one is left behind in this effort. Without urgent sociopolitical-economic responses, global suffering will escalate, jeopardizing lives and livelihoods for years to come. Immediate development responses in this crisis must be undertaken with an eye to the future.

Development trajectories in the long-term will be affected by the choices countries make now and the support they receive.

The United Nations has mobilized the full capacity of the UN system through its 131 country teams serving 162 countries and territories, to support national authorities in developing public health preparedness and response plans to the COVID-19 crisis (Wikipedia n.d.).

## 18.3  SOCIAL RESPONSE

Contemporary India's physical prosperity and cultural dynamism has often been perturbed by natural disasters, from Kashmir to Kanyakumari and from Gujarat to Arunachal Pradesh. The impact of the COVID-19 pandemic on sociopolitical economic systems and lifestyle is significant. The social response is a function of changing lifestyle. In this context, digitalization has made a landmark in the process of assessment of impact of the COVID-19 pandemic on sociopolitical and economic aspects. As such, more than ever, "New India" requires prudent use of Geo-spatial System for Resource Mapping, Distribution, and Utilization.

## 18.4  LIFESTYLE

Lifestyle refers to the way of living of a person or group. It is the consistent and integrated way of life of an individual as typified by his or her manner, attitudes, possessions, etc. The term was introduced by Austrian psychologist Alfred Adler in his book, *The Case of Miss R.*, with the meaning of "a person's basic character as established early in childhood" (Adler 1929). The broader sense of lifestyle as a "way or style of living" has been documented since 1961. Lifestyle is a combination of intangible or tangible factors. Tangible factors relate specifically to demographic variables (i.e., an individual's demographic profile), whereas intangible factors concern the psychological aspects of an individual such as personal values, preferences, and outlooks.

The most widely noticed changes in lifestyle resulting from the pandemic is the increases in online shopping. The online companies become more widespread and create very negative impact over the offline shops. This also indicates that people are using technology to meet their daily needs.

An attempt has also been made to focus on Changing Patterns and facets of the Lifestyle. From Geographical perspectives there is always a spatial relationship between the Social Response through Changing Lifestyle. The analyses depend upon the Location Based Survey (LBS) i.e., location, position, and situation of the COVID-19 sites. Besides this, location also play very vital role in developing lifestyle. Rural-urban, plain-plateau, mountainous- coastal, hot desert-glaciated, whether the region is classified on the basis of the physical, demographic, or climatic parameters, it creates very strong imprints over the lifestyle. Lifestyle may include views on politics, religion, health, intimacy, and so on.

### 18.4.1  OTHER IMPORTANT ASPECTS OF LIFESTYLE

The following are some of the aspects that play a vital role in shaping someone's lifestyle as related to Covid.

#### 18.4.1.1  Time Management

Covid-19 changed our day-to-day work habits with some people transitioning to working from home and some people losing jobs completely. Families have also had to make adjustments due to the need to care for children and elderly family members.

#### 18.4.1.2  Food Habits

Due to lockdowns and business closures, daily food habits have changed. Now the emphasis is on nutritious, simple home-cooked food, which has had a positive effect on health. Different types of supplements and herbal items have also become popular. While many of these remedies have been available for centuries, Covid has made people notice them and implement them into their daily diet.

Our ancient tradition of washing hands and face after coming from outside and before a meal as well as drinking lukewarm water have now become habits.

### 18.4.1.3 Family Relations and Space

People are now home more, which has created changes in families. People are watching TV with family, enjoying indoor games altogether, and sharing domestic chores more now. This has strengthened family bonds as well as improved quality of life.

### 18.4.1.4 Personal Space

Personal space is defined as the physical distance between two people in a social, family, or work environment. Taking this space is a sign of a healthy lifestyle. Every person needs quality time and space for relaxing. COVID-19 encroached on personal space too. People also need to socialize with others, which has been impacted by Covid. Holidays, vacations, and gatherings have changed and personal space shrinking.

### 18.4.1.5 Working Atmosphere

Many people no longer work in offices, but instead have transitioned to home offices. This has affected people socially, without the opportunity to engage in person with other people, as well as the quality of the work they produce.

### 18.4.1.6 Changing Education System

Students, teachers, and administrators have had to completely adjust to a new learning process. While virtual classrooms are adequate they cannot replace the traditional way of teaching. Students, especially who went to schools, colleges or universities for the first time, cannot conceptualize the idea of schools or colleges or universities. The campus, the classrooms, the lectures, canteens, chats and gossips with friends can not create effective imprints to their mind. That is why it is very difficult for them to visualize the concepts of educational institutions according to their level of understandings. However, India has now achieved the DESIN (Digital Education System in India).

## 18.5  HUMAN BONDS IN CRISIS

As discussed, the COVID-19 pandemic has impacted socio-economic and political systems along with social response through changing lifestyles. It has also strengthened social bonds with family and friends and created a feeling of nostalgia. People have also realized that how much kin and neighbors mean along with our sense of home and patriotism. The pandemic has greatly limited get-togethers on various occasions, even in religious gatherings (Figure 18.1) and people's ability to pay their last respects (Figure 18.1C). Chronic depression and anxiety have resulted.

India is a country with several religious belief systems, and the assembly of people at religious places plays a part in social integration and exchange of ideas (Figure 18.1D).

In fact, social response and lifestyle depend upon many beliefs about humanity and meaning in the world. In the era of the pandemic, the media contributes by reinforcing false popular beliefs or myths that create fear in people. Krishno Nimpuno 1989 has described myths and their social response with special reference to disasters such as earthquake and hurricanes.

### 18.5.1  Myths

The following are some of the myths about COVID-19 that have circulated:

**FIGURE 18.1** The social response through changing lifestyle, COVID-19.

A. Lockdown in Maharastra district from March 31, 2021. B. Strict social distancing measures implemented to contain COVID-19 spread. C. People paying last respects to a family member who died of COVID-19 at a ground for mass cremation of COVID-19 in New Delhi. D. Religious Ganga Aarti gatherings before pandemic at Banaras Ghat.

Myth 1: People in danger will panic and breakdown in hysterics or flee in panic with little consideration for the safety and security of others.

In reality, people try to stay close to relatives and neighbours and in general insist on staying near the sites of their homes even after their destruction. The first instinct of every normal person is to rescue and take care of relatives and neighbours. If fleeing, it is only in a family group, so there is no panic.

Myth 2: The helplessness of COVID-19 victims is normal; people are dazed and numb after a sudden pandemic. The disruption of daily routine creates sudden black out to everybody, but the impact varies. One who is emotionally weak and not able to deal with enormous stress easily dragged under stress and depressions. In some cases, it is also noticed that being affected to Covid is such shocking to some people that they collapses their life not for the severity of Covid but with having such trauma that they got affected to it.

Myth 3: Covid will cause long-term psychological problems and people will have severe emotional scars. As we know from the past experience that the effect of pandemic is lasting for at least 5 years.

Myth 4: Anti-social behavior, such as looting, robberies, corruptions, and women crimes, is common in pandemics. This is because losses of jobs at mass scale, collapsing of small-scale business in formal and informal sectors, retrenching of employees in multinational and national companies due to pandemic, etc. Insecurity causes family disturbances as well as personal. People who can not handle the socio-political and economic crisis have a tendency towards suicide. Along with these, poverty, untimely death due to malnutrition, domestic violence, and antisocial activities are become very common. Communities break down and morale is becoming low after the pandemic.

### 18.5.2  PHASE OF READJUSTMENT

A pandemic does not create new social behavioural patterns and social issues. Social problems that existed before the pandemic will persist after it. Post-pandemic difficulties have their origin in pre-pandemic times too. But the extremity and behavior will change to some extent. These changes are adopted in lifestyles to help people adopt to the Phase of Readjustment. This situation is termed as Equilibrium in Physical Sciences.

### 18.5.3  ADJUSTMENTS TO LIFESTYLES

The older generation is facing difficulties finding people who share their thoughts and wisdom. Before the pandemic this was easier due to listening to religious discourses, singing bhajans and hymns in unison, and being involved in social work as set by temples, gurudwaras, or churches (Figure 18.1D).

The pandemic has also brought about a sense of helplessness and inhibited the intermingling of extended families and friends thereby restricting exchange of ideas and encouraging debates and discussions, which are a part of our society .The isolation has at one side definitely brought volunteers at local levels, but feelings of sharing, disagreements, discords and then making up with friends as part of every childhood is missing. This will have a very deep impact on mental health of our young minds who will perhaps not learn the essence of forgiving, forgetting, and moving on because common classroom and play field frictions are absent now.

## 18.6  CONCLUSION

COVID-19 has forcibly prepared us for a new tomorrow. It has also taught us to appreciate that life has wonderful facets and newer avenues to walk through, which we perhaps ignored before. Positive social response through changing lifestyles is a great way for society to adopt to phases of re-adjustment. The world has faced pandemics almost every 100 years and forced us to be more systematic, meticulous, and aware of all that is happening around us.

## REFERENCES

Adler, Alfred. 1929. *The Case of Miss R.* 1st ed. New York: Taylor and Francis.
Ghebreyesus, Tedros Adhanom. 2020. "WHO Director-General's Opening Remarks at the Media Briefing on COVID-19 – 11 March 2020." World Health Organization. www.who.int/director-general/speeches/det ail/who-director-general-s-opening-remarks-at-the-media-briefing-on-covid-19---11-march-2020.
Kripalani, Krishna. 2008. *Rabindranath Tagore, a Biography*. UBS Publishers' Distributors.
Nimpuno, Krisno. 1989. Disasters and Social Response. ITC Journal, 1989-3/4, pp. 179–182
Pal, Prasanta Kumar. 2016. *Rabijibani Complete (1–9)*. 3rd ed. Ananda Publishers.
UNDP. n.d. "Accelerating-Development-Progressduring-Covid-19."
Wikipedia. n.d. "United Nations Response to the COVID-19 Pandemic." Wikipedia. Accessed June 26, 2021. https://en.m.wikipedia.org/wiki/United_Nations_response_to_the_COVID-19_pandemic.

# 19 Emerging Lifestyles in Urban India under the Regime of COVID-19 Pandemic

## An Overview of the Implications

Jagdish C. Mehta

Department of Sociology, DAV College, Chandigarh, India

## CONTENTS

## 19.1 INTRODUCTION

Pandemics are nothing new to humanity. Many notable disease outbreaks and pandemics have occurred in the history of human civilization, including Spanish flu, influenza, smallpox, cholera, swine flu Hong Kong flu, SARS, H7N9, Ebola, and Zika, all of which had devastating effects on civilization in their own peculiar way. The outbreak of the present coronavirus disease in 2019 (COVID-19), which occurred in China in December 2019, has rapidly spread all over the world, including India. The outbreak was declared a Public Health Emergency of International Concern in January 2020 and a pandemic March 11, 2020. As compared to the earlier pandemics the world witnessed, the current COVID-19 pandemic is now on the top of the list in terms of worldwide coverage. This is the first time the whole world has been affected simultaneously and struck strongly in a very short span of time (WHO, 2020). As a result, the impact of a microbial disease with global transmission and lethal potential has had a greater impact on society than at any other point in history. The pandemic of the coronavirus has turned into a human, economic, and social disaster. It has caused damage in various parts of life, including economic, social, cultural, and behavioural dimensions, in addition to disease-related health difficulties. It has led to the loss of lives and intensified human misery and toppled our lives upside down.

The Government of India ordered a nationwide lockdown for 21 days starting from March 16, 2020. Hence, the movement of the entire 1.3 billion population of India was limited as a preventive measure against the COVID-19 pandemic in India. The National Disaster Management Authority prolonged the lockdown until May 3 and then until May 31. The Ministry of Home Affairs announced that the ongoing lockdown would be further extended until June 30 in containment zones, with services resuming in a phased manner. Along with lockdown, the government is trying to limit the spread of the virus by asking citizens to adopt some behavioral changes like frequently washing their hands with running water and soap; using hand sanitizers; not shaking hands; staying at home; using face masks in public places; avoiding touching one's face, mouth, or nose; maintaining a distance of at least one meter from the next person; avoiding overcrowded places or gatherings such as places of worship and open air markets; and adopting proper hygienic, sneezing, and coughing methods.

Thus, people's lifestyles have been changed by the pandemic all around the world. Due to work-from-home instructions, the gap between personal and professional life has narrowed, and people's lives are revolving around these two due to the lockdown. People have also been pondering over a vital issue at home, namely the importance of their health and fitness (Kaur, 2020). The nature of the coronavirus and associated directions of lockdown and suggestive behavioural changes have had a bearing not only on individuals but also on other social groups, institutions, and organizations. Resultantly, new lifestyles are emerging with changing behavior patterns and practices of people.

COVID-19 has remarkably changed social behavior as well health practices such as consumption patterns including eating habits, shopping habits, recreational methods, worship places along with physical activities. Resultantly, various new lifestyle patterns have emerged that may have a long-lasting impact on individual health and society at large. Therefore, the present study intends to examine these new emerging lifestyles in socio-cultural spheres and health practices particularly in urban India brought about by the current Covid-19 pandemic and will also shed light on coping strategies for this new normal.

## 19.2   OBJECTIVES AND METHODOLOGY

The aim of this study is to explore perceived lifestyle changes after the outbreak of COVID-19 and associated preventive measures in urban India. On the one hand, the main thrust of the chapter is to examine emerging lifestyles in day-to-day social behavior and cultural practices and on the other health-related routine practices like consumer behavior in terms of eating habits and physical activities. Further, the study also focuses on the emerging digital or virtual behavior as a lifestyle that emerged as a consequence of pandemic complexities.

The study is primarily based on secondary sources, review of existing studies, and personal observation, interactive sessions with the people across groups like gender, caste, economic status, etc., along with focused group discussion among acquaintances and 80 students in B.A courses and 40 students in M.A. sociology courses at D.A.V. College, Chandigarh. The qualitative information from all these students and about their family members was collected.

The study is broadly divided in two sections. **Section I** of the study deals with the impact of the Covid-19 pandemic circumstances on the lifestyles of people in general and families in particular. **Section II** highlights the changes in health practices such as consumption patterns, the eating habits and physical fitness, which have a great bearing on overall mental and physical health of individuals.

## SECTION I

## 19.3   COVID-19 PANDEMIC: ALTERING THE LIFESTYLES OF PEOPLE

**Concept of Lifestyle:** Life is a biological construct, and living a particular way of life is a social construct. The phrase lifestyle refers to a way of life practised by individuals, groups, and nations that is shaped by geographical, economic, political, cultural, and religious contexts. The term is frequently used to refer to any of a variety of lifestyles; an individual can express himself or herself in a variety of ways, including eating habits, dressing patterns, dwelling type and location, entertainment style, leisure time pass, physical activities to preserve health, educational choices, and communication styles, among others. Lifestyle is referred to as the characteristics of inhabitants of a region at a special time and place. It includes day-to-day behaviors and functions of individuals in job, activities, fun, and diet (Farhud, 2015).

Lifestyles, according to Giddens (1991), are routinised activities incorporated into habits of dress, eating, acting, and preferred milieus for meeting others; yet, the routines followed are reflexively open to change in light of the mobile nature of self-identity (Giddenz, 2003). Lifestyle is a dynamic term as it needs to be changed with a change in circumstances and goals.

Lifestyle is the distinctive pattern of personal and social behavior characteristic of an individual or a group. Behavior includes activities involved in relationships with partners, family, relatives, friends, neighbors, and colleagues, consumption behavior, leisure, work (paid or unpaid), and civic and religious activity. Patterns of behavior are linked to values and to socio-demographic characteristics, and may involve varying necessary features of lifestyle (Veal, 1993).

However, if we separate family lifestyle from that of lifestyle in general, the family lifestyle refers to the ways that families live and coexist together on daily basis, and the habits and patterns that these people have as individuals and as part of the family unit. The way a family eats, for example, or the time they spend in physical exercise and fitness are a part of family lifestyle. Also, part of this, however, is the way they interact with each other and the activities the family does together (Miller, 2021) such as cooking, watching television, sleeping, working, travelling, and also social interaction.

Thus, lifestyle of individual, families, or any social unit consists of a pattern of behavior in terms of their regular way of life, which is changeable as per time, circumstances, and goals. Lifestyle can be expressed variously through religious practices and customs, moral values, style of dress, sexual behavior, ways of entertainment and leisure, eating habits, and various health practices. While this list is suggestive, it can extend up to any number of activities. Thus, in the present study, changes witnessed in lifestyle practices due to the Covid-19 pandemic are discussed. The following sections deal with the lifestyle changes brought by Covid-19 situations in social spheres and health practices, respectively.

## 19.4   IMPACT OF COVID-19 ON SOCIAL INDICATORS OF LIFESTYLE

The unprecedented circumstances of Covid-19 and subsequent lockdown have had a significant influence on people's lives in general and families in particular. Though primarily regarded as having a negative impact on people's lifestyles, positive changes are also noticeable in people's day-to-day activities, and a 'new normal' is emerging. Even if a large number of people do not notice any changes in their lifestyle, changes can be seen in family lifestyles, consumption patterns, physical fitness, and other individualistic behaviour patterns, particularly among some sections of society such as in urban India.

### 19.4.1   Work from Home (WFH) – A New Normal

Due to social distancing, the workplace was projected to undergo a major transformation. Employees who could work from home were encouraged to do so during the initial days of the lockdown in

India. Companies working in the IT sector in the country were early adopters of WFH. As the pandemic spread over the country, WFH became a preferred practice for employers, even for people who had never done it before. Allowing employees to work from home is expected to stay in many companies in order to reduce operating expenses while also increasing productivity (Bhattacharyya & Verma, 2020).

With little help, families have been forced to try to maintain a difficult work-family balance due to the Covid-19 pandemic. Due to the closure of schools and daycare centres, parents became completely responsible for childcare and, in some cases, schooling. While many people can work from home, not everyone has the option, and many people are facing increased financial concerns as a result of job loss. Even, it is observed that many people had to leave job in order to take care for children and elderly family members.

Families have also been forced to share responsibilities for cooking, cleaning, and childcare without the option of outside help. A positive trend is being observed in gender dynamics in which more domestic work is shared rather than being solely the woman's responsibility. The pandemic has positively influenced sensitivity among men and also their involvement in women's health.

As a result of sharing domestic tasks, some marital relationships may be reaffirmed and families strengthened. There is evidence from around the world that housework sharing leads to marital/cohabitation stability (Ruppanner et al., 2018). Due to the fragility of outside support, skills such as cooking and cleaning have been learned by all the members of the family. Moreover, the pandemic is projected to have a significant impact on the Indian economy as well since households may forego the services of cooks and maids as a precautionary measure or to save money.

### 19.4.2 Transformation in Tribulation – Changing Recreations and Reflections

Reflecting on the impact the pandemic has had on individuals' self-realization we find that in this time of lockdown, people have started to ask: what is the point of our lives? That is actually the very essence of life. In modern times, we are so preoccupied with materialistic and economic pursuits that we rarely have the opportunity to contemplate and attempt to answer these basic questions of existence. This time has allowed us to take a break from the treadmill and reflect. The members of the focused group agreed that it has been a fantastic opportunity to enjoy the peace and quiet that comes with no traffic, to hear birds chirping and other natural sounds, and to gaze at trees from the comfort of one's own home and marvel at nature's amazing beauty and intricacy. Thus, people are again coming back to the basics and creating new more peaceful lifestyles and engaging in activities such as yoga, fine arts, cooking, music, spending time with children, etc.

Isolation caused by Covid-19 has driven people to rely increasingly on electronic media devices for enjoyment or to pass the time, resulting in a rise in electronic content consumption. The amount of time spent on social media, chatting, texting, and playing video games is on the rise. Media such as Netflix, Amazon, Fox Star, and other similar apps provide massive entertainment content and are becoming increasingly popular. Online gaming has also attracted a large number of consumers who are self-isolating and staying at home, allowing numerous businesses to increase profits. If flexibility in work or work from home is practiced more widely in the aftermath of COVID-19, such consumption of electronic content is likely to continue.

### 19.4.3 Stay Away, No Visits, and Distant Greetings: A New Cultural Norm

Practicing social distancing is as much a measure to protect oneself from the virus as it is about protecting others from one's self. Phrases such as 'do your part, stay apart' and 'social distancing' are becoming ingrained in society. The government used public media to raise awareness among the public about the importance of not shaking hands for greetings, instead using various other types of greetings that do not require physical contact, like *Salaam* and *Namaste* or other greetings linked to religion, sect, community, or beliefs. This practice or new alternatives to greetings may emerge or be developed by communities and become the new norm.

### 19.4.4 Virtual Platforms: New Venues for Celebrations, Festivities, Last-rites/Prayers, and Meetings

Due to Covid-19 it was not possible to hold physical gatherings for celebrations, festivities, last rites/prayers, or meetings. The use of technology has aided the exponential expansion of virtual gatherings. Friends, coworkers, and extended family members have discovered methods to stay in touch and retain their bonds. Diverse family-based whatsapp groups have brought individuals together to discuss feelings, emotions, jokes, and to stay connected. Due to the shutdown, even marriage ceremonies, birthday celebrations, and rituals were held online using Google Meet, Webex, Zoom, Microsoft Teams, and other digital platforms and live shows were relayed via YouTube and Facebook. Organizations also used these kinds of digital platforms to hold official meetings. Even after phasing out of Covid-19, virtual events will likely continue.

### 19.4.5 Online Platforms: New Shopping Destinations

With the growing use of the internet, online shopping has become commonplace, with payment and delivery made simple. Owing to Covid-19, online shopping has become even more attractive. According to a survey by Nielsen, 95% of the consumers are afraid of coronavirus, which increased online shopping by more than 25%. The growth of online shopping is expected to remain constant even after the pandemic.

### 19.4.6 Online Learning Platforms – A New Perspective on Education

Online learning platforms are being used by educational institutes ranging from primary schools to higher education centres such as universities and colleges. Educational institutions offer online classes on platforms like Google Meet, Zoom, Web ex, Microsoft Teams, etc. The Government of India promoted digital learning through e-learning platforms like SHAGUN, DIKSHA, SWAYAM, and e-PATHSHALA. There has been a paradigm shift in the way educators deliver quality education through various online platforms. Transitioning from traditional face-to-face learning to online learning can be an entirely different experience for learners and educators. The purpose of virtual classrooms is not only to transact curriculum but also to 'exhibit care and build an effective relationship with the students' (India Today, 2020). Online learning has been observed as the best possible alternative to conventional learning. It has provided a chance to develop new and improved professional skills/knowledge in an efficient and productive way (Pravat K. Jena 2020). Such a shift to online learning could mark a turning point for the industry, changing habits in terms of how teachers teach and students learn.

## SECTION II

## 19.5   IMPACT OF COVID-19 ON HEALTH-RELATED INDICATORS OF LIFESTYLE

The COVID-19 outbreak has had a noticeable impact on health-related lifestyle patterns. Eating habits, physical activity, and other lifestyle changes can either jeopardise or improve our health. The impact of Covid-19 on the lifestyles of people depends on factors like economic situation, social status, psychological traits of their personality like attitude or approach towards life, discipline, level of health consciousness, etc. As a result, some people experienced no change or very little routinised changes in lifestyle during the COVID-19 shutdown. On the other hand, around one third of the population, particularly in metropolitan or urban areas, modified their lifestyles in healthy ways. Chopra et al. (2020) reported that most people did not change their habits (46.1%), but 16.7% and 37.2% improved them or made them worse, respectively.

### 19.5.1 Consumption Pattern Especially in Terms of Eating Habits

Consumption differs according to social class. Furthermore, consumption in India is very unique and variegated, depending on social group, product, and geography, among other factors. Meal regularity, food diversity, meal quality, meal quantity, dining traits, and other factors are all influenced positively or negatively by the circumstances of Covid-19.

Regular meal pattern as a construct is often described as an individual's eating patterns at the level of a 'meal', such as a main meal (e.g., breakfast, lunch, or dinner) or a smaller-sized meal (e.g., supper or snack) (Leech et al., 2015). Home confinement is thought to affect the regularity and frequency of meal consumption. People and families have made a variety of contradictory observations. While some people and families as a whole have improved their eating habits, others have engaged in unhealthy eating habits. During the pandemic some people have made little improvements in terms of eating meals at regular intervals on a regular basis. However, healthy eating trends have been seen, such as the consumption of protein-rich foods like pulses, eggs, and meat, as well as reduced intake of high fat, sugar, salt (HFSS) food items, especially among the younger population.

A possible reason for this difference could be more focus on homecooked food in Indian households. Consumer behaviour has shifted to homecooked meals rather than eating out as a result of differing levels of lockdown, social distance requirements, and health concerns in various countries throughout the world (Norje, 2020). Traditional cuisines and consumption behaviours, which are primarily oriented towards family, appeal to everyone in the family (including men, women, and children). People have begun to prepare their own meals in recent years, with the aim of eating healthier. More and more people are also using supplements and healthy products like herbal tea (Kadha), cinnamon, chyawanprash, dry ginger, fresh lemon, and turmeric milk.

Though outside food consumption has decreased as a result of the lockdown, consumption of HFSS and FMCG (Fast Moving Consumer Goods) foods has increased. Some research suggests that confinement increased consumption of HFSS foods, which could be due to anxiety or boredom, a drop in motivation to maintain healthy eating, or an increase in mood-driven eating. In addition to boredom, hearing or reading about COVID-19 from the media on a regular basis can be stressful.

But other factors also contribute to poor eating habits. Limited access to daily grocery shopping, for example, may lead to a reduction in fresh food consumption, particularly fruit, vegetables, and seafood, in favour of more processed meals such as convenience foods, junk foods, snacks, and ready-to-eat cereals, which tend to be high in fats, sugars, and salt. Furthermore, psychological and emotional reactions to the COVID-19 outbreak may enhance the chance of developing disordered eating habits.

A sizable proportion of people also grew grow more health concerned. Many people have changed their eating habits and turned to organic, healthful foods that are high in immunity-boosting nutrients. Immune booster foods is a growing trend among people all around the world. People are more becoming increasingly more likely to choose foods that are high in nutrition over those that are simply tasty. Many global corporations have been conducting research and development in the area of food products that are primarily designed to boost people's immunity, as demand for these products is growing by the day. Almost every major food processing company has made an investment in organic or ayurvedic processed meals or medications. Many consumers have been encouraged to pick organic plant-based, vegan, and animal-free foods as a result of COVID-19, and the demand for exotic foods has decreased significantly.

## 19.5.2   Impact on Physical Health or Fitness

Physical activity is frequently stated as vital to one's well-being since physically healthy people have a lower risk of having serious disease than physically unhealthy people (Chopra et al., 2020). Many studies have found that those with greater health or fitness are less susceptible to COVID-19 infection than those who are completely cut off from physical activity. Exercise also benefits mental health by lowering anxiety, depression, and a bad mood, as well as enhancing self-esteem and cognitive function. The imposed lockdown, which has resulted in the closure of businesses, public areas, fitness and activity centres, and overall social life, has impacted many parts of people's lives, including everyday fitness exercises, resulting in a variety of psychological disorders as well as major fitness and health concerns (Kaur et al., 2020).

Many people's physical activities have been impacted by the COVID-19 lockdown. Some people have reduced their physical activity as a result of the closing of sports facilities. Others, on the other hand, took advantage of the lockdown to begin physical activities, both because they had more time to practise and because it was one of the few reasons they could leave their houses during the lockdown.

The COVID-19 pandemic has drastically altered people's habits, with many people working from home and having minimal contact with anyone other than family members. The study by Chopra et al. (2020) revealed that average physical activity decreased in both groups, but men experienced a greater substantial decrease in their activity status than women. This may be due to the fact that women's contributions to the workforce are less affected because many stay at home.

Upper socio-economic groups have significantly better eating behavior in comparison to lower socio-economic groups, but lower socio-economic groups experienced significantly lower reduction in activity status when compared with upper socio-economic status. People who do not leave home to go to work and instead spend more time at home may have significantly reduced amounts of daily physical activity or outdoor time.

A study by Ammar et al. (2020) also found that with an increase in daily screen and sitting time, there was a substantial decline in moderate intensity aerobic exercises as well as leisure-related activities. Overall, physical inactivity was shown to be higher in men and those from higher socio-economic groups.

Due to nationwide lockdowns and home isolation, previous trends of working out in gymnasiums, playgrounds, public parks, and pools were totally cut off; this led to a fall in physical movement of those people who were used to doing daily exercises or workouts in an open environment. A reduction in engagement in physical activity at all levels coupled with increase in daily sitting and screen time due to confinement was prominently found across the literature (Ammar et al., 2020).

Despite countermeasures such as providing online 'at home physical activity courses' via different social media channels to improve overall activity at home, current findings show that individuals have not been able to effectively sustain their daily activity habits with suggested home activities (De Oliveira et al, 2020). Increased screen and sitting time followed the decrease in time spent engaged in physical activity. In our research, we found a significant increase in the number of hours (4–5 hours) spent in front of the computer for leisure or work purposes.

However, according to some reports, lockdown may have a positive impact on people's activity levels, with more people seen outside running, walking, cycling, and so on. We should be wary of interpreting this to mean that people are adopting a more active and safer lifestyles. Physical activity can be accumulated in a variety of forms. The majority of people get their 'productive minutes' by doing things like housework, walking the dog, walking/cycling to and from work, walking between train stations, and so on. All of these things are a part of people's everyday lives and lead to their total number of minutes spent exercising (www.physio-pedia.com/Physical_Activity_ and_COVID-19). But lifting heavy buckets, large water bottles, and skipping are all examples of high-intensity workouts that can be done at home.

Technology and social media play a part in raising awareness about the importance of staying healthy by monitoring physical activity on a regular basis. Wearable health devices are becoming more common, as they are appealing methods of tracking health and exercise data (Vegesna et al., 2017). During COVID various online courses in aerobics, yoga, and meditation have become available to encourage healthy living. Trainers and mentors (gurus) used social media, such as Whatsapp, Facebook, and Instagram, to inspire people to stay physically fit in order to improve immunity and protect themselves from the coronavirus.

As a result, Covid-19 has had a positive impact on certain lifestyle habits while having a negative impact on others. Despite the fact that a large proportion of people report no lifestyle changes, the percentages of people reporting favourable lifestyle changes outnumber those reporting unfavourable lifestyle changes, particularly when it comes to food quality and frequency. When it comes to leisure time, physical activity, and screen time, though, the case was the polar opposite. As a result, it is likely that COVID-19's sudden appearance caused people to rethink their lifestyle behaviours.

## 19.6 CONCLUSION

Catastrophic events like the Covid-19 pandemic bring about changes in society. Some lifestyle behaviours have been positively influenced by Covid-19, while others have been unfavourably influenced. Although a large majority of people may not notice any changes in their lifestyle, lifestyle changes are clearly visible in urban India. Maintaining social distance is likely to become habitual, with its effects going well beyond limiting the spread of the disease. The encouragement of social distancing in a variety of settings, such as the workplace, educational institutions, marketplaces, or any other public location, and in a variety of ways, is likely to make it customary in Indian culture. Changes in family life that adjust the gender relationship in favour of sharing responsibilities, saying Namaste instead of shaking hands, and virtual gatherings in place of face-to-face gatherings are all likely changes that will stay. Technology has allowed people stay connected and get healthcare in novel ways during this difficult period, and these changes are also likely permanent.

People started choosing homecooked meals and eating out less as a result of a fear of coronavirus infection, which contributed to an improvement in healthy eating and a decrease in junk food intake. There has been a considerable improvement in regular meal consumption patterns, healthy eating behaviour, and reduction in unhealthy food intake. Adverse changes in physical activity levels observed due to lack of motivation, time availability, and restricted access to parks, dance, and fitness centres have also been seen. Even though there were improvements in eating behaviors among many people, to overcome anxiety, boredom, and stress, frequent consumption of fast food items by some people was also observed. These findings may have consequences for the formulation of physical activity and dietary guidelines to help people stay healthy during the COVID-19 pandemic. After this pandemic, a new age- post-Covid in human history will begin; let us make our future race more socially concerned, healthy, fit, and wise for better quality of life. More research is needed to determine the long-term impacts of COVID-19 on lifestyle patterns.

## REFERENCES

Ammar, A., Brach, M., Trabelsi, K., Chtourou, H., Boukhris, O., & Masmoudi, L. (2020a). Effects of COVID-19 home confinement on eating behaviour and physical activity: Results of the ECLB-COVID19 international online survey. Nutrients, 12(6):1583. [PMC free article] [PubMed] [Google Scholar]

Ammar, A., Chtourou, H., Boukhris, O., Trabelsi, K., Masmoudi, L., Brach, M., et al. (2020b). COVID-19 home confinement negatively impacts social participation and life satisfaction: A worldwide multicenter study. International Journal of Environmental Research and Public Health, 17:6237. doi: 10.3390/ijerph17176237

Bhattacharyya, R., & Verma, P. (2020, April 3). Work-from-home going to stay, even after COVID-19 scare is over *The Economic Times*. https://economictimes.indiatimes.com/jobs/work-from-home-going-to-stay-even-aftercovid-scare-is-over/printarticle/74956231.cms

Borah, P. M. (2020, April 2). Couples make the most of time together during the Coronavirus lockdown. *The Hindu*. www.thehindu.com/life-and-style/how-couples-across-india-are-working-together-at-home/article31225682.ece

Chopra, S., Ranjan, P., Singh, V., Kumar, S., Arora, M., Hasan, M.S., et al. (2020). Impact of COVID-19 on lifestyle-related behaviours – A cross-sectional audit of responses from nine hundred and ninety-five participants from India. Diabetes & Metabolic Syndrome, 14(6):2021–30. 10.1016/j.dsx.2020.09.034 [PMC free article] [PubMed] [CrossRef] [Google Scholar]

De Oliveira Neto, L., Elsangedy, H.M., Tavares, V.D.D.O., Teixeira, C.V.L.S., Behm, D.G., Da Silva-Grigoletto, M.E. (2020). Training in home – home-based training during COVID-19 (SARS-COV2) pandemic: Physical exercise and behavior-based approach. RBFE, 19(2):9. [Google Scholar]

Farhud, D. (2015). Impact of life style on health. Iranian Journal of Public Health, 44 (11).

India Today (2020, April 19). Coronavirus lockdown: Here's how schools are providing seamless online educationduring COVID-19. India Today. (online). Retrieved from www.indiatoday.in/education-today/featurephilia/story/coronaviruslockdown-here-s-how-schools-are-providing-seamless-online-education-during-covid-19–1668558-2020-04–19

Jena, P.K. Challenges and opportunities created by Covid-19 for ODL: A case study of IGNOU. International Journal for Innovative Research in Multidisciplinary Filed, 6(5):217–222.

Karunathilake, K. (2020). Positive and negative impacts of Covid-19, An analysis with special reference to chanllenges on the supply chain in South Asians Countries. Journal of Social and Economic Development Institute for Social and Economic Change. Physical Activity and Covid-19 (Online). Retrieved from www.physio-pedia.com/Physical_Activity_and _COVID-19).

Kaur, H., Singh, T., Arya Y.K. & Mittal, S. (2020). Physical fitness and exercise during the COVID- 19 pandemic: A qualitative enquiry. Frontiers in Psychology. 11:590172. doi: 0.3389/fpsyg.2020.590172

Leech R.M., Worsley A., Timperio A., & McNaughton S.A. (2015). Understanding meal patterns: definitions, methodology and impact on nutrient intake and diet quality. Nutrition Research Reviews, 28(1):1–21. [PMC free article] [PubMed] [Google Scholar]

During COVID-19: Will it Change the Indian Society? Journal of Health Management, 22(2) 224–235.

Miller, B. (2021). What is family life style. Retrievef from http://www.info bloom.com.

National Center for Complementary and Integrative Health (2020). Yoga: What you need to know. (Online). Retrieved from https://nccih.nih.gov/health/yoga/introduction.htm

Nicol, G. E., Piccirillo, J. F., Mulsant, B. H., & Lenze, E. J. (2020). Action at a distance: geriatric research during a pandemic. J. Am. Geriatr. Soc. 68, 922–925.doi: 10.1111/jgs.16443

Norje, C. (2020). Economic impact of Covid-19 on the meat industry. (online). Retrieved from https://sappo.org/economicimpact-of-covid-19-on-the-meat-industry/.

Ruppanner, L., Branden, M., & Turunen, J. (2018). Does unequal housework lead to divorce? Evidence from Sweden. Sociology, 52(1): 75–94.

Sunitha. S. (2020). Covid-19: A media and entertainment sector perspective In IndiaVichar Manthan (A Peer Reviewed Journal) ISSN - 2347-9639 Vol- 8 No-3 Issue-24 August 2020 Broadcast Audience Research council of India Portal (online). Retrieved from www.barcindia.co.in/

Veal, A. (1993). The concept of lifestyle: A review. (Online). Retrieved from www.researchgate.net/publication/248996076_The_Concept_of_Lifestyle_A_Review

Vegesna, A., Tran M., Arcona S. (2017). Remote patient monitoring vis non-invasive digital technologies: A systemic review. Telemed JE Health [PMC Free Article] [google scholar]

World Health Organization (WHO) (2020). Novel coronavirus (2019-nCoV) advice for the public. (Online). Retrieved from www.who.int/emergencies/diseases/novel-coronavirus-2019/advice-for-public

# 20 Confronting Potential Role of Yoga in Combatting COVID-19 Pandemic

*Neeru Malik,[1] Rakesh Malik,[2*] and Navneet Kaur[3]*

[1]Dev Samaj College of Education sector 36-B Chandigarh, India
[2]Dr Hari Singh Gaur Vishvidyaly, A Central University,
Sagar Madhya Pradesh, India
[3]The Maharaja Bhupinder Singh Punjab Sports University, Patiala, India
*drmalikrakesh@gmail.com

## CONTENTS

## 20.1 INTRODUCTION

### 20.1.1 Yogic Philosophy and Its Multifaceted Approach in the Covid-19 Pandemic

Yoga is an Indian ancient system developed approximately 5,000 years ago. The word yoga is originated from the Sanskrit word "Yuj," which means to "join" and was first mentioned in the Rig Veda. Yoga is about attaining synchronization within the body, mind, soul, and nature. Yoga gained popularity, in particular, after the declaration of International Yoga Day by the United Nations (https://covid19.who.int/table). The beneficial role of yoga in building overall health (physical, mental, emotional, and spiritual) is well documented in the literature.

We are in the midst of the biggest health pandemic of the twenty-first century. COVID-19 has had health, economical, social, mental, and emotional effects across the world. India has had the second highest number of COVID-19 cases in the world after the United States (WHO). However, the number of mortalities is lower in India as compared to other parts of the world. Presently, India is having a recovery rate of approximately 95% (www.tribuneindia.com/news/nation/indias-covid-19-recovery-rate-crosses-95-among-highest-in-world-active-cases-below-3-4-lakh-184656). In every corner of the world people, government, social agencies, pharmaceutical companies, etc., are trying to beat this pandemic using various. means. In this unpredicted situation it is "survival of the fittest."

Part of the reason for India's good recovery rate from COVID-19 as compared to rest of the world may relate to the lifestyle of Indians, which is very closely associated with our Vedas, traditions, customs, culture, etc. During the present scenario people are often guiding each other to use *Kada*, healthy and *satvik bhojan*, steam inhalation, gold milk (turmeric milk), practice yoga and mediation for healthy living. Tillu et al. 2020 in their study also suggest Ayurveda approaches for prevention or treatment of COVID-19 such as medicated water (includes dry ginger, coriander, cinnamon, etc.), nasal oil application, gargles, and steam inhalation (Tillu et al., 2020).

COVID-19 is an infection that can lead to a variety of respiratory problems, heart issues, etc., as well as mild to moderate symptoms. People with medical conditions like diabetes, hypertension, cancer, and cardiovascular diseases develop more severe symptoms of COVID-19. Psychological problems such as depression and anxiety can also occur. Yoga may have potential in improving the immunity level of individuals along with playing a promising role in improving conditions related to respiratory illnesses. The beneficial role of yoga in reducing stress and anxiety is also supported in the literature. The soothing effect of yoga on stress and anxiety helps improve mental and spiritual health and promote immunity. Studies have also shown that yoga, which is a combination of asana, pranayama, and meditation, helps in building immunity, which results in reducing vulnerability to infections. Pranayama, which is a set of breathing exercises, has a beneficial impact on respiratory system.

## 20.1.2 CONTEXT AND METHODOLOGY

### 20.1.2.1 Context

#### 20.1.2.1.1 PATHOLOGY, SYMPTOMS, AND PREVENTION OF COVID-19

COVID-19 is a respiratory infection caused by severe acute respiratory syndrome-coronavirus 2 (Sawant et al., 2021; Nagarathna et al., 2020). The symptoms of COVID-19 range from mild to severe as shown in Figure 20.1. Moreover, it is a communicable disease in which the infection travels from one person to another very rapidly if proper preventive measures are not taken. Due to the nature of the disease and the number of COVID-19 cases the virus has spread all over the world.

The main preventive strategies for COVID-19 include avoiding touching the facial area, especially nose, eyes, and mouth, wearing masks, frequent washing of hands, and maintaining social distancing as shown in Figure 20.2 (WHO). Better immunity helps in early recovery and prevention of COVID-19 symptoms. Natural approaches like Ayurveda and yoga might help in effectively combatting COVID-19.

According to the Bhagvadgeeta:

"योगस्थः कुरु कर्माणि संग त्यक्त्वा धनंजय ।
सिद्धियसिद्धियोः समो भूत्वा समत्वं योग उच्यते" ॥ ॥2.48॥ (Gita, Chapter 2)

which means yoga helps to maintain the mind-body equilibrium. Moreover, yoga helps to maintain balance between our behavior, thoughts, and emotions (www.yogapoint.com,www.yugalsarkar.com)

Moreover, according to Maharishi Patanjali who developed the concept of Ashtanga yoga defines yoga as "Yogah Chitta Vritti Nirodhah."

"योगः चतित्त-वृत्ति निरोधः" (Patanjali's Yoga-Sutra, 2nd Sutra)

which means that through yoga one can control the activities of the mind, which includes various feelings, emotions, sensations, and thoughts (www.yogapoint.com).

The multidimensional benefits of yoga are well documented in the literature (Nayak, 2011; Taneja, 2014; Desveaux, 2015; Bernstein, 2014). Yoga helps to maintain synchronization among the different body systems and helps in proper functioning of different body organs. Yoga not only

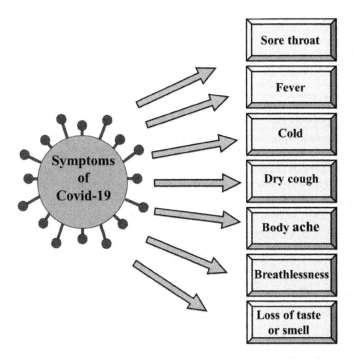

**FIGURE 20.1**    Symptoms of COVID-19 (Center for Disease Control and Prevention, 2020) (redrawn).

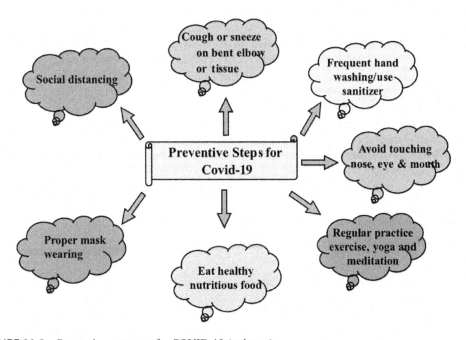

**FIGURE 20.2**    Preventive measures for COVID-19 (redrawn).

benefits the body physically but also at mental and psychological levels. It also works effectively on psychosomatic disorders (http://indianyoga.org/wp-content/uploads/2020/03/v9-issue1-article10. pdf) by maintaining homeostasis balance of the body.

### 20.1.3 Methodology

The present chapter reviews the current literature, looking at the potential role of yoga in managing the symptoms associated with COVID-19. Moreover, the therapeutic role of yoga in managing post-COVID-19 complications and its potential to improve physical, mental, and emotional issues are also elucidated. Various databases were searched like Google Scholar, Pub med, ERIC, and Directory of Open Access Journals (DOAJ) for literature related to the keywords COVID-19, yoga, COVID-19 symptoms, prevention, exercise, physical health, mental health, and immunity. The systematic reviews, cross-sectional studies, pilot studies, randomized controlled trials (RCT), and non-RCTs related to yoga and Covid-19 were examined. The literature covers COVID-19 symptoms and precautions, and the role of yoga in managing, regulating, and controlling them.

### 20.1.4 Yoga as an Essential Tool in the Covid-19 Pandemic

#### 20.1.4.1 Yoga as an Immunity Booster

The immune system in our body plays a key role in fighting various infections. A dysregulated immune system is one of the key features (Nagendra, 2020) in severe conditions of COVID-19 cases. The white blood cells (WBCs) present in our blood act as a defence system in our body, and fight against unwanted toxic substances, infections, and viruses. A strong immune system helps build a defense against COVID-19. Moreover, a strong immune system results in better recovery for mild to moderate cases of COVID-19. Yoga is a non-pharmacological method, and has the potential to boost immunity.

A study by Naoroibam et al. (2016) on HIV-1 infected patients were given one month of yoga treatment. After one month of yoga practice the HIV patients had significant improvements in CD 34 immune cells. Another study reported significant association between peaceful mind and immunity of cells through yoga intervention (Kamei et al., 2001). Available evidence shows that yoga might have a potential role in building and maintaining immunity level. Yoga practices are also beneficial in reducing pro-inflammatory markers. A study showed a noteworthy decline in pro-inflammatory markers like interleukine-6 (IL-6), interleukine-1 (IL-1), and TNF –alpha after yogic practices (Falkenberg et al., 2018). It is also well known that COVID-19 severely affects people suffering from other medical conditions and yoga may be an alternative therapy (Falkenberg et al., 2018). The findings from studies suggest that yoga might enhance the immunity (Nagarathna, 2019) of people infected with COVID-19 or people who are at high risk of developing COVID-19.

#### 20.1.4.2 Yoga for Physical Health

व्यायामात् लभते स्वास्थ्यं दीर्घायुष्यं बलं सुखं।
आरोग्यं परमं भाग्यं स्वास्थ्यं सर्वार्थसाधनम् ॥

The meaning of above Sanskrit *shloka* is that regular exercise helps in bringing good health, longer age, more strength, and happiness.

The WHO also recommended being physically active. As discussed earlier COVID-19 severely affects people with pre-existing health conditions. Yoga and physical activity can improve pre-existing health conditions, since physical exercise helps in managing blood pressure, body weight, cardiovascular diseases, and diabetes (www.who.int/news-room/campaigns/ connecting-the-world-to-combat-coronavirus/healthyathome/healthyathome---physical-activity? gclid=Cj0KCQiAvbiBBhD-ARIsAGM48bx0ZLMDlBCcKOXU7-wsAmfrZpX8YJSs9b3k 3aMS8z-KQTp1rDodZ3EaAseSEALw_wcB). Physical inactivity is a major risk for many non-communicable diseases, and can make people more vulnerable to severe COVID-19 if contracted.

Due to lockdown many people had more time for daily physical activity like yoga. The various types of physical postures help in maintaining the homeostasis of the body. The practice of yoga not only offers physical fitness but also helps in rejuvenating the functioning of various organs along with better mind alertness. Yoga also includes breathing practices like pranayama, which greatly help in improving the respiratory functioning of the individual (Santaella et al., 2011). Pranayama not only helps to establish control of breathing, but also helps to improve lung functionality up to alveolar level.

### 20.1.4.3   Yoga for Psychological Health

According to Yoga Vashitha

"मनःप्रशमनोपायो योग इत्यभिधीयते॥" (www.yogapoint.com)

The above *shloka* denotes that yoga is a perfect way to eliminate different types of problems related to the mind. It helps keep the mind calm, poised, and peaceful. Control and balance may further help in controlling body activities (www.yogapoint.com).

The COVID-19 pandemic has had a severe impact on mental health due to lockdowns, illnesses/deaths, economic issues, etc. Yoga is considered to be a very potent therapy for improving mental and psychological health. Poor mental health is associated with various respiratory problems (Maxwell et al., 2015). The beneficial effect of yoga on mental stress and depression has been reported in published studies (Benavides and Caballero, 2009; Sharma et al., 2008). Yoga aids various psychological and cognitive variables like perceived stress, anxiety and depression, focus, etc. (Kaur et al., 2020). Yoga includes breathing and meditative practices, which results in better lung efficiency and reduced mental issues, respectively.

### 20.1.5   ROLE OF YOGA IN CURRENT PANDEMIC

As the saying goes, "prevention is better than cure." This saying is very appropriate in the current COVID-19 pandemic. It is imperative for individuals to incorporate daily physical activity, breathing exercises, and meditation to deal with the stress of living with COVID. In hatha yoga pradipika the importance of pranayama was elaborated as:

"प्राणायामेन युक्तेन, सर्वरोगक्षयो भवेत् ।" (blog.practicalsanskrit.com). This particular *shloka* suggests that by doing pranayama in a proper way all illness, sickness, or disease may be cured (blog.practicalsanskrit.com).

The Ministry of AYUSH proposed the Common Yoga Protocol, which is a combination of asana, pranayama, and meditative techniques. The Common Yoga Protocol is suitable for all age groups and has an ameliorating effect on the psychological, neuronal, and immunity of people who practice (Ministry of AYUSH, 2014). The Common Yoga Protocol is a standarised protocol that can help during the pandemic since yoga practices have beneficial impact on stress, anxiety, and depression along with physical benefits (Ofei-Dodoo et al., 2020). Nagarathna et al. (2020) report the various yoga protocols according to different age groups for preventing and controlling COVID-19. The researches based on these yoga protocol depicted interesting result and shows the recovery in COVID-19 patients (Nagarathna et al., 2020). Moreover, yoga is an economical and non-pharmacological approach, and can be done easily at home or in the workplace without side effects. The trend of online yoga classes has increased during the pandemic due to social distancing and lockdowns. However, more research is required to fully understand the role and efficacy of yoga in COVID-19. However, the existing literature gives us hope that yoga can positively impact physical and mental health along with boosting immunity.

## 20.2  CONCLUSION

The COVID-19 pandemic has brought various types of challenges to mankind. We have been forced to adjust ourselves to the "new normal." The principles of yoga, which aim for unification of body, mind, and soul, can help us effectively handle the COVID-19 challenge. Yoga and Ayurveda practices might play an adjuvant role in dealing with COVID-19 and in managing post COVID-19 complications.

## REFERENCES

Benavides, Sandra, and Joshua Caballero. "Ashtanga yoga for children and adolescents for weight management and psychological well being: an uncontrolled open pilot study." *Complementary Therapies in Clinical Practice* 15, no. 2 (2009): 110–114.

Bernstein, Adam M., Judi Bar, Jane Pernotto Ehrman, Mladen Golubic, and Michael F. Roizen. "Yoga in the management of overweight and obesity." *American Journal of Lifestyle Medicine* 8, no. 1 (2014): 33–41.

Desveaux, Laura, Annemarie Lee, Roger Goldstein, and Dina Brooks. "Yoga in the management of chronic disease." *Medical Care* 53, no. 7 (2015): 653–661.

Falkenberg, R. I., C. Eising, and M. L. Peters. "Yoga and immune system functioning: A systematic review of randomized controlled trials." *Journal of Behavioral Medicine* 41, no. 4 (2018): 467–482.

Kamei, T., Yoshitaka Toriumi, H. Kimura, and Keishin Kimura. "Correlation between alpha rhythms and natural killer cell activity during yogic respiratory exercise." *Stress and Health: Journal of the International Society for the Investigation of Stress* 17, no. 3 (2001): 141–145.

Kaur, Navneet, Neeru Malik, Deepali Mathur, Surinder Pal, Rakesh Malik, and Sumit Rana. "Mindfulness and yoga halt the conversion of pre-diabetic rural women into diabetics-a pilot study." *Age (Years)* 42: 8–18.

Maxwell, Lizzie, Bruce Barrett, Joseph Chase, Roger Brown, and Tola Ewers. "Self-reported mental health predicts acute respiratory infection." *WMJ: Official Publication of the State Medical Society of Wisconsin* 114, no. 3 (2015): 100.

Ministry of AYUSH, I., Common Yoga Protocol. 2014.

Nagarathna, R., H. R. Nagendra, and Vijaya Majumdar. "A perspective on yoga as a preventive strategy for coronavirus disease 2019." *International Journal of Yoga* 13, no. 2 (2020): 89.

Nagendra, H. R. "Yoga for COVID-19." *International Journal of Yoga* 13, no. 2 (2020): 87.

Naoroibam, Rosy, Kashinath G. Metri, Hemant Bhargav, R. Nagaratna, and H. R. Nagendra. "Effect of Integrated Yoga (IY) on psychological states and CD4 counts of HIV-1 infected patients: A randomized controlled pilot study." *International Journal of Yoga* 9, no. 1 (2016): 57.

Nayak, Arun Kumar. "Effects of yoga on mental and physical health." *IJPESH* 3, no. 5 (2016): 268–270.

Ofei-Dodoo, Samuel, Anna Cleland-Leighton, Kari Nilsen, Jacob L. Cloward, and Eastin Casey. "Impact of a mindfulness-based, workplace group yoga intervention on burnout, self-care, and compassion in health care professionals: A pilot study." *Journal of Occupational and Environmental Medicine* 62, no. 8 (2020): 581–587.

Santaella, Danilo F., Cesar R.S. Devesa, Marcos R. Rojo, Marcelo B.P. Amato, Luciano F. Drager, Karina R. Casali, Nicola Montano, and Geraldo Lorenzi-Filho. "Yoga respiratory training improves respiratory function and cardiac sympathovagal balance in elderly subjects: a randomised controlled trial." *BMJ Open* 1, no. 1 (2011): e000085.

Sawant, Ranjeet S., Bharat D. Zinjurke, and Sandeep V. Binorkar. "Preventive aspect of ayurveda and yoga towards newly emerging disease COVID-19." *Journal of Complementary and Integrative Medicine* (2021).

Sharma, Kanupriya, Akshay Anand, and Raj Kumar. "The role of Yoga in working from home during the COVID-19 global lockdown." *Work* Preprint (2020): 1–7.

Sharma, R. A. T. N. A., Nidhi Gupta, and Ramesh Lal Bijlani. "Effect of yoga based lifestyle intervention on subjective well-being." *Indian Journal of Physiology and Pharmacology* 52, no. 2 (2008): 123–131.

Taneja, Davendra Kumar. "Yoga and health." *Indian Journal of Community Medicine: Official Publication of Indian Association of Preventive & Social Medicine* 39, no. 2 (2014): 68.

Tillu, Girish, Sarika Chaturvedi, Arvind Chopra, and Bhushan Patwardhan. "Public health approach of ayurveda and yoga for COVID-19 prophylaxis." *The Journal of Alternative and Complementary Medicine* 26, no. 5 (2020): 360–364.

WEB REFERENCES

Bhagavad Gita Chapter 2 Verse 48. www.yugalsarkar.com/bhagwad-gita-chapter-2-shlok-48 (accessed 10 December 10, 2020).

Centers for Disease Control and Prevention. 2020. Symptoms of COVID-19. www.cdc.gov/coronavirus/2019-ncov/symptoms-testing/symptoms.html (accessed December 13, 2020).

https://blog.practicalsanskrit.com/2011/01/importance-of-proper-breathing.html (accessed December 16, 2020).

www.yogapoint.com/info/what-is-yoga.htm (accessed December 17, 2020).

Singh, P. 2020. Psychosomatic management through Yoga. *International Journal of Yoga and Allied Sciences,* no. 9 (January): 68–73 http://indianyoga.org/wp-content/uploads/2020/03/v9-issue1-article10.pdf (accessed December 12, 2020).

The Tribune. 2020. www.tribuneindia.com/news/nation/indias-covid-19-recovery-rate-crosses-95-among-highest-in-world-active-cases-below-3-4-lakh-184656 (accessed December 15, 2020).

World Health Organization. 2020. #Healthy at Home – Physical Activity. www.who.int/news-room/campaigns/connecting-the-world-to-combat-coronavirus/healthyathome/healthyathome---physical-activity?gclid=Cj0KCQiAvbiBBhD-ARIsAGM48bx0ZLMDlBCcKOXU7-wsAmfrZpX8YJSs9b3k3aMS8z-KQTp1rDodZ3EaAseSEALw_wcB (accessed December 11, 2020).

World Health Organization. 2020. Coronavirus Disease (COVID-19) Advice for the Public. www.who.int/emergencies/diseases/novel-coronavirus-2019/advice-for-public (accessed December 13, 2020).

World Health Organization. 2020. Situation by Region, Country, Territory & Area. https://covid19.who.int/table (accessed December 2, 2020).

# 21 Lifestyle Communication and Environment Sustainability

## An Analytical Study from Consumption to Information Dissemination during Covid-19

*Mahendra Kumar Padhy*

Department of Mass Communication, Babasaheb Bhimrao
Ambedkar University, Lucknow, India

## CONTENTS

DOI: 10.1201/9781003358916-21

## 21.1  INTRODUCTION

Ever since the end of the Second World War there has been a common concern for macro-level development among development theorists and policymakers. The last six decades have witnessed a remarkable shift in the meaning and focus of development. While sustainable development is a concept that has been around for two decades it is considered as the latest trend in human development at local as well as global levels. This chapter deals with the issue of sustainable development and the role of the media. Is the role of sustainable development the same or different in relation to previous trends of development? If it is different, to what extent? Why and how? It becomes pertinent to ask these questions for two reasons: One, the scope of sustainable development is broader than earlier notions of development. While earlier development was the concern of a few economists, policymakers, and administrators, sustainable development has become everybody's concern. Two, it is pertinent to find out the implications of the tremendous changes that have taken place in communication in relation to sustainable development (Dahlan, 1989).

There is no medicine to cure diseases like Covid-19; prevention is key. Similarly, informed and conscious citizens can take steps to avert danger to the environment and promote safety. Therefore, lifestyle communication can play an important role in sustainable development by creating awareness, educating the people, translating the technical knowledge into the people's language, conscientizing the people, and facilitating people's expressions and interaction at the grassroots level (Brown, 1992).

## 21.2  CONCEPTUAL BACKGROUND

Sustainable development as an emerging discipline occupies a pivotal place in every aspect of human life today. Sustainable development has become the concern of media academicians, economists, ecologists, administrators, lawyers, communication experts, environmentalists, human rights activists, and non-governmental organizations (NGOs). In other words, it has become everybody's cup of tea.

The World Commission on Environment and Development (WCED) defined sustainable development as the development that meets the needs of the present without compromising the ability of future generations to meet their own needs (Brundtland, 1987).

It is an assumption that sustainable development is a coin with an important obligation on each side. One side is the alleviation of poverty and the other the protection of the environment. Sustainable development is very much linked with the personal involvement and active participation of people. It is a holistic concept that can be on the global, national, local, and individual scale. The media is an intervening variable without which the materialization of different goals of sustainable development is not possible. Therefore, the media has a key role in facilitating the participation of people relating to sustainable development.

## 21.3  RESEARCH OBJECTIVES

As sustainable development is the most recent and modern trend of social development, the broad aim of this study is to analyze the role of participatory development communication for sustainable development and to recommend a suitable communication strategy for sustainable development. The following are some of the important objectives earmarked for this study:

1. To identify the role and implications of sustainable social development during Covid-19.
2. To study and examine the developmental programs of government towards the alleviation of poverty nationally during Covid-19.

3. To study and examine the role of participatory development media in communicating lifestyles for the protection of biodiversity in the Koraput district of Odisha.

## 21.4   RESEARCH DISCUSSION

Statistical Analysis and Interpretation of the Data.

### 21.4.1   CHI-SQUARE TEST OF INDEPENDENCE

The Chi-Square Test of Independence is used to test if two categorical variables are associated.

The null hypothesis (H) and alternative hypothesis (H1) is defined as:

H0: (Variable 1) is independent of (Variable 2)

H1: (Variable 1) is not independent of (Variable 2)

Test Statistic: The test statistic for the Chi-Square Test Independence is denoted as X2, and is computed as, $\chi2=\sum(oij-eij)2/eij$, where oij is the observed cell count in the ith row and jth column of the table; and eij is the expected cell count in the ith row and jth column of the table, computed as eij= row i total *col j total/grand total.

The calculated X2 value is then compared to the critical value from the X2 distribution table with degrees of freedom df = (R – 1)(C – 1) and chosen confidence level. Since the p-value is greater than our chosen significance level ($\alpha$=0.05), we do not reject the null hypothesis. Rather, we conclude that there is no association between variables.

#### 21.4.1.1   AWARENESS OF GOVERNMENT SCHEMES FOR POVERTY ALLEVIATION (MNREGA, IRDP, AND SGRY) AND OF SUSTAINABLE DEVELOPMENT

The cross-tabulation and count for awareness of government schemes for poverty alleviation (Mahatma Gandhi National Rural Employment Guarantee Act 2005 (MNREGA), Integrated Rural Development Program (IRDP), and Sampoorna Grameen Rozgar Yojana (SGRY)) and of sustainable development are shown in Tables 21.1 and 21.2.

The value of the test statistic is 12.256. The corresponding p-value of the test statistic is p= 0.000. Since the p-value is less than our chosen significance level ($\alpha$=0.05), we reject the null hypothesis. We conclude that there is an association between awareness of government schemes for poverty alleviation (MNREGA, IRDP, and SGRY) and awareness of sustainable development (ozone layer, global warming).

**TABLE 21.1**

**Cross Tabulation of Awareness of Government Schemes for Poverty Alleviation (MNREGA, IRDP, and SGRY) and of Sustainable Development**

|  |  | Awareness of Sustainable Development (Ozone Layer, Global Warming) | | |
|---|---|---|---|---|
|  |  | No | Yes | Total |
| Awareness of Government Schemes for Poverty Alleviation | No | 86 | 29 | 115 |
| (MNREGA, IRDP and SGRY) | Yes | 104 | 86 | 190 |
|  | Total | 190 | 115 | 305 |

**TABLE 21.2**
**Cross-Tabulation Count**

| Chi-Square Tests | Value | df | Asymp. Sig. (2-sided) | Exact Sig. (2-sided) | Exact Sig. (1-sided) |
|---|---|---|---|---|---|
| Pearson Chi-Square | 12.256[a] | 1 | .000 | | |
| Continuity Correction[b] | 11.417 | 1 | .001 | | |
| Likelihood Ratio | 12.615 | 1 | .000 | | |
| Fisher's Exact Test | | | | .001 | .000 |
| Linear-by-Linear | | 16.1 | | 000 | |
| Association | | 12.2 | | | |
| N of Valid Cases | | 305 | | | |

*Note:* [a] 0 cells (.0%) have expected count less than 5. The minimum expected count is 43.36. [b] Computed only for a 2×2 table.

**TABLE 21.3**
**Cross Tabulation of Awareness of Government Schemes for Poverty Alleviation (MNREGA, IRDP, and SGRY) and Awareness of Media (Television, Newspaper, Internet)**

| | | Awareness of Media (Television, Newspaper, Internet) | | |
|---|---|---|---|---|
| | | No | Yes | Total |
| Awareness of Government Schemes for Poverty Alleviation (MNREGA, IRDP and SGRY) | No | 5 | 110 | 115 |
| | Yes | 6 | 184 | 190 |
| Total | | 11 | 294 | 305 |

**TABLE 21.4**
**Cross-Tabulation Count**

| Chi-Square Tests | Value | df | Asymp. Sig. (2-sided) | Exact Sig. (2-sided) | Exact Sig. (1-sided) |
|---|---|---|---|---|---|
| Pearson Chi-Square | | | 1 .589 | .292[a] | |
| Continuity Correction[b] .050 | | | 1 .823 | | |
| Likelihood Ratio | | | 1 .593 | .286 | |
| Fisher's Exact Test | | | | .753 | .403 |
| Linear-by-Linear Association | | | .291 1 | .590 | |

*Note:* [a] 1 cell (25.0%) has expected count less than 5. The minimum expected count is 4.15. [b] Computed only for a 2×2 table.

### 21.4.1.2 Awareness of Government Schemes for Poverty Alleviation (MNREGA, IRDP, and SGRY) and Awareness of Media (Television, Newspaper, Internet)

The cross-tabulation and count for awareness of government schemes for poverty alleviation (Mahatma Gandhi National Rural Employment Guarantee Act 2005 (MNREGA), Integrated Rural Development Program (IRDP), and Sampoorna Grameen Rozgar Yojana (SGRY)) and of media (television, newspaper, internet) are shown in Tables 21.3 and 21.4.

**TABLE 21.5**
**Media Literacy Campaign on Developmental Schemes**

| Chi-Square Tests | Value | df | Asymp. Sig. (2-sided) |
|---|---|---|---|
| Pearson Chi-Square | 19.095[a] | 5 | .002 |
| Likelihood Ratio | 19.973 | 5 | .001 |
| Linear-by-Linear Association | 1.453 | 1 | .228 |
| N of Valid Cases | 305 | | |

*Note:* a. 4 cells (33.3%) have expected count less than 5. The minimum expected count is .31.

The value of the test statistic is 0.292. The corresponding p-value of the test statistic is p= .589. Since the p-value is greater than our chosen significance level ($\alpha$=0.05), we do not reject the null hypothesis. We conclude that there is no association between awareness of government schemes for poverty alleviation (MNREGA, IRDP, and SGRY) and awareness of media (television, newspaper, internet).

### 21.4.1.3 Awareness of Health-Related Issues and Sources of the Developmental Scheme for Poverty Alleviation and Food Security

The value of test statistic is 19.095. The corresponding p-value of the test statistic is p= .002. Since the p-value is less than our chosen significance level ($\alpha$=0.05), we reject the null hypothesis, and we conclude that there is an association between awareness of health-related issues and source of motivation of the developmental scheme for poverty alleviation and food security.

### 21.4.2 CORRELATIONS

The sample correlation coefficient between two variables x and y is denoted r or r can be computed as, $r_{xy}$=cov(x,y)/var(x) $\sqrt{}$ var(y) where cov(x, y) is the sample covariance of x and y, var(x) is the sample variance of x, and var(y) is the sample variance of y.

Correlation can take on any value in the range [–1, 1]. The sign of the correlation coefficient indicates the direction of the relationship, while the magnitude of the correlation (how close it is to –1 or +1) indicates the strength of the relationship:

y –1: perfectly negative linear relationship
y 0: no relationship
y +1: perfectly positive linear relationship

Hypothesis: The null hypothesis (H0) and alternative hypothesis (H1) is defined as:
H0: $\rho$ = 0 ("the population correlation coefficient is 0; there is no association")
H1: $\rho \neq 0$ (the population correlation coefficient is not 0; a non-zero correlation could exist)
where P is the population correlation coefficient.

### 21.4.2.1 Awareness of Government Schemes for Poverty Alleviation (MNREGA, IRDP, and SGRY) and Awareness of Sustainable Development (Ozone Layer, Global Warming)

The correlation of awareness of government schemes for poverty alleviation (MNREGA, IRDP, and SGRY) itself is r=1 and the number of observations is n=305.

Correlation of awareness of government schemes for poverty alleviation (MNREGA, IRDP, and SGRY) and awareness of sustainable development (ozone layer, global warming) is r= 0.200 and the number of observations is n=305.

**TABLE 21.6**
**Media Strategy and Sustainable Development**

| | | Awareness of Government Schemes for Poverty Alleviation (MNREGA, IRDP and SGRY) | Awareness of Sustainable Development (Ozone Layer, Global Warming) |
|---|---|---|---|
| Awareness of Government Schemes for Poverty Alleviation (MNREGA, IRDP, and SGRY) | Pearson Correlation | 1 | .200 |
| | Sig. (2-tailed) | | .000 |
| | N | 305 | 305 |
| Awareness of Sustainable Development (Ozone layer, Global warming) | Pearson Correlation | .200 | 1 |
| | Sig. (2-tailed) | .000 | |
| | N | 305 | 305 |

**TABLE 21.7**
**Awareness of Government Schemes for Poverty Alleviation**

| | | Awareness of Government Schemes for Poverty Alleviation (MNREGA IRDP, and SGRY | Awareness of Media (Television, Newspaper, Internet) |
|---|---|---|---|
| Awareness of Government Schemes for Poverty Alleviation (MNREGA, IRDP, and SGRY) | Pearson Correlation | 1 | 031 |
| | Sig. (2-tailed) | | 591 |
| | N | 305 | 305 |
| Awareness of Media (Television, Newspaper, Internet) | Pearson Correlation | .031 | 1 |
| | Sig. (2-tailed) | 591 | |
| | N | 305 | 305 |

Correlation of awareness of sustainable development (ozone layer, global warming) itself is $r=1$ and the number of observations is $n=305$.

Awareness of government schemes for poverty alleviation (MNREGA, IRDP, and SGRY) and awareness of sustainable development (ozone layer, global warming) have a statistically significant linear relationship ($p<.001$). The direction of the relationship is positive (i.e., positively correlated).

### 21.4.2.2  Awareness of Government Schemes for Poverty Alleviation (MNREGA, IRDP, and SGRY) and Awareness of Media (Television, Newspaper, Internet)

Correlation of awareness of government schemes for poverty alleviation (MNREGA, IRDP, and SGRY) itself is $r=1$ and the number of observations is $n=305$.

Correlation of awareness of government schemes for poverty alleviation (MNREGA, IRDP, and SGRY) and awareness of media (television, newspaper, internet) is $r=0.031$ and the number of observations is $n=305$.

**TABLE 21.8**
**Media Literacy and Health Communication**

| | | Awareness of Government Schemes for Poverty Alleviation (MNREGA, IRDP, and SGRY) | Awareness of Media (Television, Newspaper, Internet) |
|---|---|---|---|
| Awareness of Government Schemes for Poverty Alleviation (MNREGA, IRDP, and SGRY) | Pearson Correlation | 1 | 031 |
| | Sig. (2-tailed) | | 591 |
| | N | 305 | 305 |
| Awareness of Media (Television, Newspaper, Internet) | Pearson Correlation | .207 | 1 |
| | Sig. (2-tailed) | 000 | |
| | N | 305 | 305 |

Correlation of awareness of media (television, newspaper, internet) itself is r=1 and the number of observations is n=305.

Awareness of government schemes for poverty alleviation (MNREGA, IRDP, and SGRY) and awareness of media (television, newspaper, internet) have a statistically significant linear relationship (p<.001). The direction of the relationship is positive (i.e., positively correlated).

### 21.4.2.3 Awareness of Government Schemes for Poverty Alleviation (MNREGA, IRDP, and SGRY) and Awareness of Health-Related Issues

Correlation of awareness of government schemes for poverty alleviation (MNREGA, IRDP, and SGRY) itself is r=1 and the number of observations is n=305.

Correlation of awareness of government schemes for poverty alleviation (MNREGA, IRDP, and SGRY) and awareness of health-related issues is r=0.207 and the number of observations is n=305.

Correlation of awareness of health-related issues itself is r=1 and the number of observations is n=305.

Awareness of Government Schemes for Poverty Alleviation (MNREGA, IRDP, and SGRY) and Awareness of Health-related issues have a statistically significant linear relationship (p<.001). The direction of the relationship is positive (i.e., positively correlated).

### 21.4.3 T-TEST

T-test: The test statistic for a One Sample t-Test is denoted t, and is calculated using the following formula:

$$t = \bar{x} - \mu/s\bar{x}$$

where $s\bar{x} = s/\sqrt{n}$
$\mu$ = Proposed constant for the population mean $\bar{x}$ = Sample mean
n = Sample size (i.e., number of observations)
s = Sample standard deviation

sx⁻ = Estimated standard error of the mean (s/sqrt (n)).

The calculated t value is then compared to the critical t value from the t distribution table with degrees of freedom df = n − 1 and chosen confidence level. If the calculated t value > critical t value, then we reject the null hypothesis.

H0: μ = x ("the sample mean is equal to the [proposed] population mean")
H1: μ ≠ x ("the sample mean is not equal to the [proposed] population mean")

where μ is a constant proposed for the population mean and x is the sample mean.

Awareness of government schemes for poverty alleviation (MNREGA, IRDP, and SGRY, Pradhan Mantri Awas Yojana (PMAY)):
The value of test statistic is t = 20.612 and mean difference is 0.573.
The degree of freedom is 304.

**TABLE 21.9**
**One-Sample Statistical Test of Government Development Schemes**

|  | N | Mean | Std. Deviation | Std. Error Mean |
|---|---|---|---|---|
| Awareness of Government Schemes for Poverty Alleviation (MNREGA, IRDP, and SGRY) | 305 | .62 | .485 | .028 |
| Awareness of Sustainable Development (Ozone layer, Global warming) | 305 | .38 | .485 | .028 |
| Awareness of Media (Television, Newspaper, Internet) | 305 | .96 | .187 | .011 |
| Awareness of Health-related Issues | 305 | .69 | .464 | .027 |
| Source of Motivation of the Developmental Scheme for Poverty Alleviation and Food Security | 305 | 1.20 | 1.058 | .061 |

**TABLE 21.10**
**One-Sample Test on MNREGA Scheme**

|  | t | df | Sig. (2-tailed) | Mean | Difference | 95% Confidence Interval of the Difference | Lower | Upper |
|---|---|---|---|---|---|---|---|---|
| Awareness of Government Schemes for Poverty Alleviation (MNREGA, IRDP and SGRY) | 20.612 | 304 | .000 |  | .573 |  | .52 | .63 |
| Awareness of Sustainable Development (Ozone layer, Global Warming) | 11.766 | 304 | .000 |  | .327 |  | .27 | .38 |
| Awareness of Media (Television, Newspaper, Internet) | 85.464 | 304 | .000 |  | .914 |  | .89 | .93 |
| Awareness Health-related Issues | 24.040 | 304 | .000 |  | .639 |  | .59 | .69 |
| Source of Motivation of the Developmental Scheme for Poverty Alleviation and Food Security | 18.931 | 304 | 000 |  | 1.147 |  | 1.03 | 1.27 |

Awareness of sustainable development (ozone layer, global warming):
    The value of test statistic is t = 11.766 and mean difference is 0.327.
    The degree of freedom is 304.

Awareness of media (television, newspaper, internet)
    The value of test statistic is t = 85.464 and mean difference is 0.914.
    The degree of freedom is 304.

Awareness of health-related issues
    The value of test statistic is t = 24.040 and mean difference is 0.634.
    The degree of freedom is 304.

Source of motivation of the developmental scheme for poverty alleviation and food security
    The value of test statistic is t = 18.931 and mean difference is 1.147.
    The degree of freedom is 304.

Since $p<0.001$, we reject the null hypothesis that the sample mean is equal to the hypothesized population mean and conclude that the mean (awareness of government schemes for poverty alleviation (MNREGA, IRDP, and SGRY), awareness of sustainable development (ozone layer, global warming), awareness of media (television, newspaper, internet), awareness of health-related issues, and source of motivation of the developmental scheme for poverty alleviation and food security) of the sample is significantly different that the average of overall population (Adams, 1995).

### 21.4.4   REGRESSION

Regression: Linear regression is a linear approach to modelling the relationship between dependent variables and independent variables.

A simple linear regression model is:

$$Y = \beta$$

$$0 + \beta1 \, X$$

where Y is a dependent variable, and X is an independent variable. The terms $\beta0$ and $\beta1$ are the parameters of the model. $\beta0$ is an intercept parameter, and $\beta1$ is a slop parameter.

Hypothesis: If there is a significant linear relationship between the independent variable X and the dependent variable Y, the slope will be not equal to zero. The null hypothesis (H0) states that slope is equal to zero and the alternative hypothesis (H1) states that the slope is not equal to zero:

$$H0: \beta1 = 0$$

$$H1: \beta1 \neq 0$$

If the p-value is less than significance level ($\alpha=0.05$) then we say the model is significant.

### 21.4.4.1   Awareness of Government Schemes for Poverty Alleviation (MNREGA, IRDP, and SGRY)

Dependent variable: Sex (male and female).
Independent variable: Awareness of government schemes for poverty alleviation (MNREGA, IRDP, and SGRY).

The coefficient of determination is 0.188; therefore 18.8% of the variation in sex data is explained by awareness of government schemes for poverty alleviation (MNREGA, IRDP, and SGRY).

Since the p-value (0.000) is less than ($\alpha=0.05$) we reject the null hypothesis. At the $\alpha=0.05$ level of significance there exists enough evidence to conclude that the slope of population regression line is not zero.

**TABLE 21.11**
**Model Summary[b]**

| Model | R | R Square | Adjusted R Square | Std. Error of the Estimate | Durbin-Watson |
|---|---|---|---|---|---|
| 1 | .434[a] | .188 | .186 | .452 | 2.092 |

*Note:* [a] Predictors: (Constant), awareness of government schemes for poverty alleviation (MNREGA, IRDP, and SGRY); [b] dependent variable: Sex.

**TABLE 21.12**
**ANOVA[a] Test on Beneficiaries of Poverty Alleviation Schemes**

| | Sum of Squares | df | Mean | Square F | Sig. |
|---|---|---|---|---|---|
| Model Regression | 14.352 | 1 | 14.352 | 70.266 | .000[b] |
| 1 Residual | 61.890 | 303 | .204 | | |
| Total | 76.243 | 304 | | | |

*Note:* [a] Dependent variable: Sex; [b] predictors: (Constant), awareness of government schemes for poverty alleviation (MNREGA, IRDP, and SGRY).

**TABLE 21.13**
**Confidence Level and Coefficients of IRDP and SGRY Schemes[a]**

| Model | Unstandardized Coefficients Beta | Standardized Coefficients | t | Sig. | 95% Confidence Interval for B | |
|---|---|---|---|---|---|---|
| B | | | | | Lower Bound | Upper Bound |
| Std. Error | 1.774 | 042 | 42.091 | 000 | 1.691 | |
| 1.857 (Constant) | | | | | | |
| 1 Awareness about Government Alleviation for Poverty Schemes | −.448 −.343 | 053 | −8.383.434 | 000 | −553 | |

*Note:* [a] Dependent variable: Sex.

We are 95% confident that the slope of the true regression line is somewhere between −0.553 and −0.349.

### 21.4.4.2  Awareness of Sustainable Development (Ozone Layer, Global Warming)

Dependent variable: Sex (male and female).
Independent variable: Awareness of Sustainable Development (Ozone Layer, Global Warming).
The coefficient of determination is 0.012; therefore 1.2% of the variation in sex data is explained by awareness of sustainable development (ozone layer, global warming). Since the p-value (0.061) is greater than ($\alpha$=0.05) we do not reject the null hypothesis.

We are 95% confident that the slope of the true regression line is somewhere between (−0.227 and 0.005).

**TABLE 21.14**
**Model Summary**[b]

| Model | R | R Square | Adjusted R Square | Std. Error of the Durbin-Watson Estimate |
|---|---|---|---|---|
| 1 | .107[a] | .012 | .008 | .499 1.995 |

*Note:* [a] Predictors: (Constant), awareness of sustainable development (ozone layer, global warming); [b] dependent variable: Sex.

**TABLE 21.15**
**ANOVA**[a]

| | Sum of Squares | df | Mean | Square F | Sig. |
|---|---|---|---|---|---|
| Model Regression | .879 | 1 | .879 | 3.533 | .061b |
| 1 Residual | 75.364 | 303 | .249 | | |
| Total | 76.243 | 304 | | | |

*Note:* [a] Dependent variable: Sex; [b] predictors: (Constant), awareness of sustainable development (ozone layer, global warming).

**TABLE 21.16**
**Coefficients**[a]

| Model B | Unstructured Coefficients | Standardized Coefficients | Significance Level A | Sig. B | 95% Confidence Interval for B |
|---|---|---|---|---|---|
| B | Beta | – | – | Lower Bound | Upper Bound |
| Std. Error .857(Constant) | 1.537 | .036 | 42.476 | 42.476 | 1.466 |
| Awareness of Sustainable Development (Ozone Layer, Global Warming) | 1,608 | –.111 | –059 | –.107 | –.1.880 |

*Note:* [a] Dependent variable: Sex.

### 21.4.4.3 Awareness of Media (Television, Newspaper, Internet)

Dependent variable: Sex (male and female).
Independent variable: Awareness of media (television, newspaper, internet).

## 21.5 CONCLUSION

To conclude, the study reveals that awareness of the schemes is a crucial factor for the proper utilization of the schemes – MGNREGA, SGRY, Swarnjayanti Gram Swarozgar Yojana (SGSY), and IRDP – to provide employment during this global pandemic. While all the beneficiaries of these

**TABLE 21.17**
**Model Summary[b]**

| Model | R | R Square | Adjusted R Square | Std. Error of the Durbin-Watson Estimate |
|---|---|---|---|---|
| 1 | .090[a] | .008 | .005 | .500 1.976 |

*Note:* [a] Predictors: (Constant), awareness of media (television, newspaper, internet); [b] dependent variable: Sex.

**TABLE 21.18**
**ANOVA[a]**

| | Sum of Squares | df | Mean Square | F | Sig. |
|---|---|---|---|---|---|
| Model Regression | .615 | 1 | .615 | 2.465 | .117[b] |
| 1 Residual | 75.627 | 303 | .250 | | |
| Total | 76.243 | 304 | | | |

*Note:* [a] Dependent variable: Sex; [b] predictors: (Constant), awareness of media (television, newspaper, internet).

schemes have awareness of these schemes, 36% of the MGNREGA and SGRY target beneficiaries and 25% of the SGSY and IRDP target beneficiaries have not used these schemes and do not have awareness of the schemes. Those who have awareness of these schemes acquired this awareness through interpersonal and group communication channels: Government officials, the sarpanch (head of the village), friends, and family members. Interestingly, no single respondent mentioned mass media as a source of information (Abraham, 1980). Moreover, 54% of MGNREGA and SGRY non-beneficiaries could not use the scheme, lack of awareness being one of the reasons. Where there is more awareness and initiative on the part of the beneficiaries, the effective use of the scheme is found to be greater. For example, in MGNREGA and SGRY beneficiaries in Nandapur and SGSY and IRDP beneficiaries in Pottangi awareness and initiative have been found to be higher. These respective schemes also have been utilized properly. In the case of the majority of the MGNREGA and SGRY beneficiaries, there is some improvement in their financial condition, as every month they are saving some small amount of money. However, some beneficiaries are not able to pay money every month, as their husbands continue to drink alcohol. Therefore, there is an indispensable need to create awareness and educate the people (both men and women) about the importance of regular savings and about the disadvantages of drinking, in the case of SGSY and IRDP asset-giving schemes.

This study reveals that government officials have given adequate choice to the target beneficiaries to choose the asset they need. But since they select the beneficiaries, they have sanctioned the loans to those who are better off and thus show bias in favour of rich applicants instead of providing opportunity to the poor for whom this scheme is meant. Thus, on the one hand in Nandapur, beneficiaries failed to utilize the schemes, and on the other hand, in Pottangi, beneficiaries could utilize the scheme, but they already had a certain level of affluence. Overall the assets have been given to a very limited number of people, which is a shortcoming of the scheme and which limited its impact. Though for 80% of the respondents there has been no improvement in their quality of life, they feel that there is now a change in their outlook, which is a commendable achievement of the scheme.

Interpersonal and group communications among the members of the MGNREGA and Unnat Bharat Abhiyan groups have contributed significantly to this change.

## REFERENCES

Abraham, F. (1980). *Perspective on Modernization: Toward a General Theory of Third World Development*. Washington, DC: University Press of America.

Adams, W.M. (1995). Green Development Theory? In J. Crush (ed.). *Power of Development*. London: Routledge. pp. 87–99.

Brown, L.R., Flavin, C., and Postal, S. (1992). From Growth to Sustainable Development. In R.J.A. Goodland, H.E. Daly, and S.E. Serafy (Eds.), *Population, Technology, and Lifestyle: The Transition to Sustainability*. Washington, DC: Island Press, pp. 119–127.

Brundtland, G.H. (1987). *Report of the World Commission on Environment and Development: Our Common Future*.

Dahlan Alwi, M. (1989). The Environmental Approach to Mass Media Coverage. *Media Asia*, 16(4), 219–222.

# 22 Changes in Diet Pattern during Covid Pandemic

*Komal Singh*

Department of Food and Nutrition, BBAU Satellite Campus,
Amethi, India

## CONTENTS

## 22.1 INTRODUCTION

COVID-19 is caused by a specific coronavirus (SARS-CoV 2). Coronaviruses primarily cause enzootic infections in animals but are capable of transferring to humans and causing infections. Covid affects the upper and the lower parts of the respiratory system. The most common symptoms are fever, dry cough, and tiredness, and less common symptoms include aches and pains, sore throat, diarrhea, conjunctivitis, headache, loss of taste or smell, a rash on the skin, and discoloration of the fingers or toes.

The first case of coronavirus was reported in Wuhan China on December 31, 2019. In just a period of 18 months, over 150 million cases and over 3 million deaths worldwide were reported. It has affected 220 countries all over the world and the subsequent lockdowns resulted in loneliness, social breakup, and lethargy among people leading to low-quality food intake.

## 22.2 IMPACT OF COVID-19 ON HEALTH

COVID-19 has greatly impacted the general well-being of humans. In Nigeria, the lockdown period was associated with an increase in food prices, a decrease in dietary diversification, elevated GAD symptoms, and disrupted diet and consumption patterns. There were low levels of physical activity and weight gain during the lockdown period, thus increasing the risk of becoming overweight and obesity. Studies incorporating participants of different socioeconomic status are warranted for conclusive results.

DOI: 10.1201/9781003358916-22

The COVID-19 blockade has had an impact on the dietary habits and nutritional patterns of the countries affected. As a result, some differences in the type and duration of home confinement, cultural and social tendencies of the countries, age of the sample analysed, and level of previous obesity have resulted in different conclusions. In particular, it has been reported in Spain that the diet followed during the lockdown had a higher energy intake and lower nutritional quality than pre-COVID-19 eating patterns (Batlle-Bayer et al. 2020). Furthermore, a 6% increase in daily intake was observed during the COVID-19 home confinement when compared to the same period in 2019. In this vein, a study conducted in Poland's population revealed that people ate and snacked more during the lockdown, with these tendencies being more common in overweight and obese people (Sidor & Rzymski 2020). One possible explanation for this increase in energy consumption could be that staying at home, indoors, and working remotely had a direct impact on daily food habits, resulting in an increase in energy ingestion and a craving for comfort food due to boredom and stress (Muscogiuri et al. 2020). Governments must consider that lifestyle changes that alter diet patterns and food quality, such as increased consumption of high caloric foods and decreased consumption of healthy foods such as vegetables and fruits, may increase the risk of chronic diseases (Rundle et al. 2020).

In terms of food composition, COVID-19 has altered the frequency and amount of consumption of some products (Rundle et al. 2020; Rodriguez-Besteiro et al. 2021). These changes supported the hypothesis that staying at home and social distancing have a negative impact on adherence to healthy dietary patterns (Rodriguez-Besteiro et al. 2021). Specifically, a previous study found that during the lockdown, Spanish people consumed fewer beverages, slightly more eggs and red meat, and consumed significantly more plant-based foods such as nuts, pasta, and vegetables. As compared to the previous year 2019 (Rodriguez-Besteiro et al. 2021). Despite these changes, red meat consumption remained higher than the recommended dietary guidelines, while plant-based food consumption remained below the recommended range (Rodriguez-Besteiro et al. 2021). During the COVID-19 lockdown, Chinese people's dietary patterns changed, with a decrease in the frequency of intake of fresh vegetables and fruit, rice, poultry, meat, and soybean products (Jia et al. 2021).

Interestingly, another study (Ruiz-Roso et al. 2020) found that COVID-19 confinement resulted in a healthy dietary change in the adolescent population of Brazil, Chile, Colombia, Spain, and Italy, increasing the frequency of ingestion of vegetables, legumes, and fruit during the lockdown. As a result, the proportion of adolescents consuming the recommended weekly servings of legumes and fruits during confinement increased by 8% and 7.7%, respectively, when compared to the pre-confinement period. These pattern changes may be justified for a variety of reasons. First, since the beginning of confinement, the sale of legumes and fruits has increased, and second, people have more time to cook at home (Ruiz-Roso et al. 2020). As a result, controversial findings have been reported in recent studies that examine the effect of lockdown on dietary patterns. As a result, some potential factors may have an impact on the reported results. For example, studies examined some populations with innate eating patterns. Furthermore, each country's government imposed different levels of lockdown, with varying degrees of severity and restrictions on population. These factors could explain the inconsistent patterns observed in the studies reported (Opichka, Smith & Levine 2019; Ruiz-Roso et al. 2020).

Surprisingly, among all groups affected by the COVID-19 lockdown, overweight and obese people have the worst dietary and lifestyle habits. These populations are known to exhibit more disruptive eating behaviours, such as food consumption without hunger and frequent overeating (Opichka et al. 2019). In this way, overweight and obese individuals reported eating and snacking more during home confinement, which could be explained by the prolonged stay at home with often unlimited food access. Furthermore, individuals with a higher BMI consumed less fruit, vegetables, and legumes during the lockdown and consumed more dairy, meat, and fast foods (Błaszczyk-Bębenek et al. 2020).

In 2019, nearly 750 million people, or nearly one in every ten people worldwide, faced severe food insecurity. According to preliminary estimates, the COVID-19 pandemic could add between 83 million and 132 million people to the global undernourished population in 2020, depending on the economic growth scenario (FAO, IFAD, UNICEF, WFP, & WHO 2020).

Prolonged malnutrition and lack of micronutrients affect the cytokine response and the transport of immune cells. Chronic inflammation and malnutrition affect immune response. However, not only is undernutrition a problem. Patients with obesity (excess fat storage) have demonstrated chronic low-grade inflammation in systemic circulation with higher concentrations of inflammatory markers (Bourke, Berkley & Prendergast 2016).

Good quality food and nutrition may not cure the virus, but it can help to boost the immune system of the body against the disease. All over the world doctors are recommending supplements of vitamins and minerals to boost the immune system.

It has seen that the people who have a high fat and sugar intake are obese and are more at risk of the virus. The high rate of consumption of diets high in saturated fats, sugars, and refined carbohydrates (collectively called the Western diet, WD) worldwide contribute to the prevalence of obesity and type 2 diabetes, and could place these populations at an increased risk for severe COVID-19 pathology and mortality.

Death and hunger have been double challenges for marginalised people. To face these challenges, the world must unite and think and act in such a way that no one is left behind (UN slogan), and everyone's life matters. This is the time to generate the philosophy of survival (Bhandari et al. 2019a, 2019b), live and let us help others to live, and Bashudhaiva Kutumbakam (the entire world is our home and all living beings are our relatives) (Bhandari et al. 2019) – in the case of humans, the threat of death and hunger is a problem of humanity, so we must face challenges and overcome them together, no matter where people live and whatever their circumstances are. All people are affected by the pandemic. However, the disease has a greater impact on the elderly (Butler & Barrientos 2020). In addition to the classic COVID-19 symptoms, older patients with COVID-19 exhibit geriatric frailty symptoms such as confusion, walking impairments, and a high mortality rate (Karlsson et al. 2021).

Malnutrition or co-malnutrition is a risk in older patients with COVID-19 disease. SARS-CoV-2 attacks mucosal epithelium and causes gastrointestinal symptoms, worsening the nutritional status of elderly patients, according to Huang et al. (2020). Many of the identified risk factors for COVID-19 viral infections and deaths are linked to nutritional status and specific essential nutrients. It is well understood that essential nutrients play an important role in maintaining the immune system's normal functions (Richardson & Lovegrove 2021).

## 22.3 NUTRIENTS TO BOOST THE IMMUNE SYSTEM

Existing evidence suggests that the only sustainable way to survive in the current situation is to strengthen the immune system. An adequate intake of zinc, iron, and vitamins A, B 12, B6, C, and E is essential for the maintenance of immune function.

### 22.3.1 Vitamin A

Vitamin A shows cytoprotective action, anti-viral activity, anti-inflammatory effects, and immunity-based immunomodulation against SARS-CoV. The anti-coronavirus benefits may be the dual efficacy of a nutrient agent and bioactive compound to treat complex disease by synergistically modulating all presumptive multitargets and multipathways.

All-trans-retinol and retinyl esters or β-carotene are different known forms of vitamin A that are present in different foods (Blomhoff & Blomhoff 2006; Mora, Iwata, & Von Andrian 2008; Moise et al. 2007). Common metabolites of vitamin A have been identified to have important effects on adaptive immune responses (Mora et al. 2008). Retinoic acid effects on the immune system were reported as enhanced cytotoxicity (Dennert & Lotan 1978), T-cell proliferation (Dennert & Lotan 1978), stimulation of Il-2 secretion, signaling in T cells (Dennert & Lotan 1978), and T cell functions. Vitamin A deficiency in mice has been also reported as leading to defects in TH-cell activity (Carman, Smith, & Hayes 1989).

Vitamin A given to malnourished children against measles in developing countries has resulted in considerable decreases in deaths due to viral infection (Munasir et al. 1995; Semba et al. 1997). Vitamin A supplementation also has an enhancing effect on antibody titers in humans (Semba 1999). Vitamin A has also been reported to augment function of leukocytes, reduce susceptibility to infective factors like carcinogenic factors, and increase integrity of mucosal membrane, which acts as a first line of defence against pathogen penetration (Alpert 2017).

Retinoic acid, which is known as an important metabolite of vitamin A, is produced in gastro-intestinal dendritic cells and has some unique functions. It has been reported to play several roles in the immune system including potent stimulation activity on TH2-cells, considerable differentiation inhibition activity of TH17-cells, and development induction activity in TReg-cell (Mora, Iwata, & Von Andrian 2008).

As noted by the above mentioned points vitamin A can be considered as an important nutrient in strengthening the immune system and during viral infections including COVID-19. Unfortunately, at the current time there is no study on the effects of vitamin A supplements in prevention and/or healing of COVID-19 infection. The effects of vitamin A in COVID-19 infected patients and/or animals can be considered as a novel topic for future studies.

## 22.3.2 VITAMIN B COMPLEX

Members of B complex include thiamine (B1), riboflavin (B2), niacin (B3), pantothenic acid (B5), pyridoxine (B6), biotin (B7), folic acid (B9), and cobalamins (B12). Several members of the vitamin B complex are used in our body for prevention of diseases by boosting the immune system and promoting a defensive role for better health.

Vitamin forms falling under this group of vitamin B complex play a vital role by functioning as antioxidants in the body, thus improving the efficacy of the immune response. Members of the vitamin B complex that have such roles in immune response efficacy include vitamin B6, B9, and B12 (Li et al. 2020).

Vitamin B12 against bacterial infection and induced shock subsequent to bacterial lipopolysaccharide was studied in mice (Toyosawa et al. 2004). Intravenous administration of pure vitamin B2 is recommended as a potent choice in prevention and/or treatment of any kind of sepsis (Toyosawa et al. 2004). It was suggested that molecules that are synthesized subsequent to metabolism of B vitamins by bacteria may result in activation of a T cell called mucosa-associated invariant T (MAIT) cells, which are a unique class of immune cells (Mikkelsen et al. 2017).

It was reported that vitamin B6 deficiency has an impact on immune cells and their specific activities is more than other members of the vitamin B family, which may be due to more sensitivity of immune cells to vitamin B6 deficiency (Chandra & Sudhakaran 1990).

Vitamin B12 also plays an important role in the development of the immune system as it is responsible for cell division and production. White blood cells cannot grow and expand when B12 is in short supply. Healthy older immune competent adults with low serum concentration of vitamin B12 had impaired antibody responses to the pneumococcal polysaccharide vaccine (Aslam et al. 2017).

Folic acid is a water-soluble vitamin and a synthetic source of folate. Regulation of immune responses and inhibition of homocystein-induced NF-κB activation in cultured human monocytes were reported as important roles of vitamin B9 (Mikkelsen et al. 2017).

### 22.3.3 Vitamin C

Since its first synthesis in 1933, vitamin C has been proposed as a treatment for respiratory infections (Haworth & Hirst 1933). Linus Pauling, the Nobel laureate, concluded from early randomised controlled trials (RCTs) that vitamin C prevented and relieved the common cold, and thus popularised its use in the 1970s (Pauling 1970, 1971). It is thought to have an antiviral effect via direct virucidal activity and increased interferon production, as well as effector mechanisms in both the innate and adaptive immune systems (Thomas & Holt 1978; Hemilä 1998; Webb & Villamor 2007). The release of reactive oxygen species (ROS) from activated phagocytes is part of the host response to viruses and bacteria. This, paradoxically, has been shown to be harmful to host cells and, in some cases, is implicated in viral and bacterial pathogenesis (Goode & Webster 1993; Peterhans 1997; Akaike, Suga & Maeda 1998). Animal studies support vitamin C's beneficial role in reducing the occurrence and severity of bacterial and viral infections (Hemilä 2006). Positive effects are an increased resistance to infection of chick embryo tracheal organ cultures and protection of broiler chicks against avian coronavirus (Atherton, Kratzing & Fisher 1978; Davelaar & Van Den Bos 1992; Ziegler-Heitbrock et al. 1993).

Vitamin C ascorbic acid is an antioxidant and scavenger of free radicals and improves immunity. Vitamin C is a heat-sensitive chemical. Extra cooking reduces the vitamin C content in food so it is better to eat raw or half-cooked food. Many fruits and vegetable like blackberries, strawberries, oranges, lemons, tomatoes, broccoli, mustards, green kale, and cauliflower are good sources of vitamin C. Beetroot, kiwi, guava, blackberries, papaya, lemon, lime, and even potato also provide a good quantity of vitamin C.

### 22.3.4 Vitamin D

Vitamin D is unique: it is a pro-hormone released in the skin during sunlight exposure (UVB radiation at 290–315 nm), typically with lower quantities obtained from food. Many people, especially those living in the northern latitudes (such as the United Kingdom, Ireland, Northern Europe, Canada, and the northern parts of the United States, Northern India, and China) have low vitamin D status, particularly in winter (Rahman & Idid 2020).

The severity of vitamin D lack is separated into mild (25-hydroxyvitamin D less than 20 ng/mL), moderate (25-hydroxyvitamin D less than 10 ng/mL), and severe (25-hydroxyvitamin D less than 5 ng/mL).

Low vitamin D status may be exacerbated during this COVID19 crisis (e.g., due to indoor living and hence reduced sun exposure), and anyone who is self-isolating with limited access to sunlight is advised to take a vitamin D supplement according to their government's recommendations for the general population (i.e., 400 IU/day for the UK and 600 IU/day for the United States (800IU for >70 years) and the EU) (Ebadi & Montano-Loza 2002).

Vitamin D agonist, calcitriol, exhibits a protective effect against acute lung injury by modulating the expression of members of the rennin-angiotensin system such as ACE2 in lung tissue, supporting the role of vitamin D deficiency as a pathogenic factor in COVID-19.

### 22.3.5 Vitamin E

Vitamin E is a fat-soluble vitamin with several forms, but alpha-tocopherol is the only one used by the human body. Its main role is to act as an antioxidant. Vitamin E is essential to preserving older people's physical well-being and their immunity. Vitamin E is an effective antioxidant that can provide protection against various pathogens, bacteria, and viruses. To get the daily dosage of vitamin E, soaked almonds, peanut butter, sunflower seeds, and even hazelnuts should be eaten (Traber & Sies 1996).

Vitamin E is a major fat-soluble antioxidant that scavenges peroxyl radicals and terminates the oxidation of polyunsaturated fatty acids (PUFAs). In the presence of vitamin E, peroxyl radicals react with α-tocopherol instead of lipid hydroperoxide, the chain reaction of peroxyl radical production is stopped, and further oxidation of PUFAs in the membrane is prevented (Traber 2007).

Vitamin E has been shown to have a beneficial impact in enhancing the production of T-cell immune synapse and activating a T-cell response. Vitamin E supplementation restores interleukin2 (IL-2) development when administered to humans, which increases the overall functioning of T-cell proliferation and the immune system. Thus, increasing sources of vitamin E in the diet of the elderly may be advantageous for their immune function, may provide resistance to infection, and may decrease morbidity because of infections. As the elderly population is more vulnerable to infection, investigating vitamin E for possible health benefits against COVID-19 would be advantageous for improving T-cell proliferation and the overall functioning of the immune system (Arshad et al. 2020).

## 22.3.6 Trace Minerals

### 22.3.6.1 Zinc

Zinc is classified as a trace mineral.

The body only requires a small amount. It aids in the growth and defence of immune cells. Zinc has been shown to inhibit viral entry, polyprotein production, and viral RdRp activity.

Zinc is a key factor in the activity and proliferation of neutrophils, NK cells, macrophages, T and B lymphocytes, and the production of cytokines by immune cells (Te Velthuis et al. 2010).

Free intracellular $Zn2+$ is required for extravasation to the site of infection, as well as the uptake and killing of microorganisms by neutrophils (Lewis, Meydani & Wu 2019). Te Velthuis and colleagues demonstrated that $Zn2+$ inhibited in vitro RdRp activity using recombinant SARS-CoV nsp12. They also reported that $Zn2+$ specifically inhibited SARS-CoV RdRp elongation and template binding. $Zn2+$ has previously been shown to inhibit the proteolytic processing of replicase polyproteins (Rahman & Idid 2020). Overall, the available data on zinc's immunomodulation effect suggested further research to investigate the clinical evidence and prove the conditional statement of the association with chloroquine (CQ) and hydroxychloroquine (HCQ). Several theoretical studies have suggested that increasing intracellular $Zn2+$ concentration by CQ may modulate the antiviral effect against SARS-CoV-2. In this regard, zinc supplementation in the absence of CQ may produce similar effects without the negative side effects of CQ treatment (Te Velthuis et al. 2010).

### 22.3.6.2 Iron

Iron is one of the most important minerals for every age group. People suffering from iron deficiency are more at risk of developing acute respiratory tract infections.

Reduced levels of iron result in thymus atrophy and affect the activity of naïve T lymphocytes like zinc. T lymphocyte proliferation decreases up to 50% to 60% due to a low iron level (Arshad et al. 2020).

Evidence also suggests that iron chelators can exhibit an antiviral effect on HIV through the elevation of intracellular iron efflux and increasing iron exporter ferroportin expression. Although to date little is known about iron regulation in COVID-19 patients, it could be deduced from other viral infections that iron chelation might be an alternative beneficial adjuvant in treating COVID-19.

### 22.3.6.3 Selenium

Selenium has an important effect on both innate and acquired immunity. Selenium enhances the function of T lymphocyte and B lymphocyte and also increases the activity of the natural killer cells (Khan et al. 2017). A study found that selenium supplementation improved immune function in the human body. The common sources of selenium are fish, meat, eggs, and nuts. Supplementation of selenium also has some adverse effects on the body (Tayyib et al. 2020).

The amount of trace elements present in food varies according to the geographical differences of the soil. In this regard, soils in different regions of China have been reported to have the highest and lowest selenium levels in the world. Zhang et al. (2020) found that infected patients from areas with high selenium levels were more likely to recover from COVID-19 (Fernández-Quintela et al. 2020).

The immune system responds to the SARS-CoV-2 via a cytokine storm and hyper inflammation, which itself leads to further multi-organ damage and, in the worst scenario, to death (Song et al. 2020). Yet, it is a fact that people consuming a well-balanced diet are healthier, with a strong immune system, and present a reduced risk of chronic illness and getting infectious diseases (Song et al. 2020). Furthermore, the immune system is always active, which is accompanied by an increased metabolism, requiring energy sources, substrates for biosynthesis, and regulatory molecules. These energy sources, substrates, and regulatory molecules are ultimately derived from the diet. Hence an adequate supply of a wide range of nutrients is essential to support the immune system to function optimally (Calder 2020). B complex vitamins are important in viral and bacterial infections (Yoshii et al. 2019). Vitamin C is thought to be antiviral (Cerullo et al. 2020). Vitamins E and D, as well as zinc, have been found to be essential for the immune system and are currently being studied in relation to COVID-19. High protein consumption (>15%) may be a top priority because it stimulates immunoglobulin production and may cause antiviral activity (Ng et al. 2015). Furthermore, recent studies suggest that individuals should consume fruit, vegetables, legumes, nuts, whole grains, unsaturated fats, white meats, and fish as part of a regular meal. Fruit juice, tea, and coffee can also be consumed cautiously because sweetened fruit juices, fruit juice concentrates, syrups, fizzy drinks, and still drinks must be avoided (Chowdhury et al. 2020). Saturated fat, red meat, more than 5g of salt per day, and industrially processed foods should all be avoided (Herrera-Peco et al. 2021).

## 22.4 CONCLUSION

COVID-19 is a highly deadly contagious virus. It has affected the whole world and had social economic impacts that have affected nutritional intake. A proper diet can help to ensure that the body is in the strongest possible state to battle the virus. An adequate intake of zinc, iron, and vitamins A, B12, B6, C, and E is essential for the maintenance of immune function. Dietary guidelines by experts will help to control COVID-19.

## REFERENCES

Akaike, T., Suga, M., Maeda, H. (1998). Free radicals in viral pathogenesis: Molecular mechanisms involving superoxide and NO. *Proceedings of the Society of Experimental Biology and Medicine*, 217(1), 64–73.

Alpert, P. T. (2017). The role of vitamins and minerals on the immune system. *Home Health Care Management & Practice*, 29(3), 199–202.

Arshad, M. S., Khan, U., Sadiq, A., Khalid, W., Hussain, M., Yasmeen, A., Asghar, Z., Rehana, H. (2020). Coronavirus disease (COVID-19) and immunity booster green foods: A mini review. *Food Science & Nutrition*, 8(8):3971–6.

Aslam, M. F., Majeed, S., Aslam, S., Irfan, J. A. (2017). Vitamins: Key role players in boosting up immune response – A mini review. *Vitamin Miner*, 6(1), 2376–1318.

Atherton, J. G., Kratzing, C. C., Fisher, A. (1978). The effect of ascorbic acid on infection of chick-embryo ciliated tracheal organ cultures by coronavirus. *Archives of Virology*, 56, 195–199.

Batlle-Bayer, L., Aldaco, R., Bala, A., Puig, R., Laso, J., Margallo, M., Vázquez-Rowe, I., Antó, J. M., Fullana-i-Palmer, P. (2020). Environmental and nutritional impacts of dietary changes in Spain during the COVID-19 lockdown. *Science of the Total Environment*, 748, 141410.

Bhandari, M. P. (2019a). "Vasudhaiva Kutumbakam" – The entire world is our home, and all living beings are our relatives. Why we need to worry about climate change, with reference to pollution problems in the major cities of India, Nepal, Bangladesh, and Pakistan. *Advances in Agricultural and Environmental Science*, 2(1), 8–35. doi:10.30881/aaeoa.00019

Bhandari, M. P. (2019b). Live and let other live-the harmony with nature /living beings-in reference to sustainable development (SD)-is contemporary world's economic and social phenomena is favorable for the sustainability of the planet in reference to India, Nepal, Bangladesh, and Pakistan? *Advances in Agricultural and Environmental Science*, 2(2), 37–57. doi:10.30881/aaeoa.00020

Błaszczyk-Bębenek, E., Jagielski, P., Bolesławska, I., Jagielska, A., Nitsch-Osuch, A., Kawalec, P. (2020). Nutrition behaviors in Polish adults before and during COVID-19 lockdown. *Nutrients*, 12, 3084.

Blomhoff, R., Blomhoff, H. K. (2006). Overview of retinoid metabolism and function. *Journal of Neurobiology*, 66(7), 606–630.

Bourke, C. D., Berkley, J. A., Prendergast, A. J. (2016). Immune dysfunction as a cause and consequence of malnutrition. *Trends in Immunology*, 37(6), 386–398.

Butler, M. J., Barrientos, R. M. (2020). The impact of nutrition on COVID-19 susceptibility and long-term consequences. *Brain, Behavior, and Immunity*, 87, 53–54.

Calder, P. C. (2020). Nutrition, immunity and COVID-19. *BMJ Nutrition, Prevention & Health*, 3, 74–92.

Carman, J. A., Smith, S. M., Hayes, C. E. (1989). Characterization of a helper T lymphocyte defect in vitamin A-deficient mice. *Journal of Immunology* (Baltimore, Md.: 1950), 142(2), 388–393.

Cerullo, G., Negro, M., Parimbelli, M., Pecoraro, M., Perna, S., Liguori, G., Rondanelli, M., Cena, H., D'Antona, G. (2020). The long history of vitamin C: From prevention of the common cold to potential aid in the treatment of COVID-19. *Frontiers of Immunology*, 11, 574029.

Chandra, R. K., Sudhakaran L. (1990). Regulation of immune responses by vitamin B6. *Annals of the New York Academy of Sciences*, 585, 404–423. doi:10.1111/j.1749-6632.1990.tb28073.x

Chowdhury, M. A., Hossain, N., Kashem, M. A., Shahid, M. A., Alam, A. (2020). Immune response in COVID-19: A review. *Journal of Infection and Public Health*, 13(11), 1619–1629.

Davelaar, F. G., Van Den Bos, J. (1992). Ascorbic acid and infectious bronchitis infections in broilers. *Avian Pathology*, 21, 581–589. doi:10.4060/ca9692en

Dennert, G., Lotan, R. (1978). Effects of retinoic acid on the immune system: stimulation of T killer cell induction. European *Journal of Immunology*, 8(1), 23–29.

Ebadi M., Montano-Loza A. J. (2020) Perspective: Improving vitamin D status in the management of COVID-19. *European Journal of Clinical Nutrition*, 1–4.

FAO, IFAD, UNICEF, WFP and WHO. (2020), The State of Food Security and Nutrition in the World 2020. Transforming food systems for affordable healthy diets. Rome: FAO.

Fernández-Quintela, A., Milton-Laskibar, I., Trepiana, J., Gómez-Zorita, S., Kajarabille, N., Léniz, A., González, M., Portillo, M. P. (2020). Key aspects in nutritional management of COVID-19 patients. *Journal of Clinical Medicine*, 9(8), 2589.

Goode, H. F., Webster, N. R. (1993). Free radicals and antioxidants in sepsis. *Critical Care Medicine*, 21, 1770–1776.

Haworth, W. N., Hirst, E. L. (1933). Synthesis of ascorbic acid. *Journal of the Society of Chemical Industry (London)*, 52, 645–647.

Hemilä, H. (1998). Vitamin C and infectious diseases. In *Vitamin C*, pp. 73–85. Milan: Springer.

Herrera-Peco, I., Jiménez-Gómez, B., Peña-Deudero, J. J., De Gracia, E. B. (2021). Comments on nutritional recommendations for CoVID-19 quarantine. *European Journal of Clinical Nutrition*, 1–2.

Huang, C., Wang, Y., Li, X., Ren, L., Zhao, J., Hu, Y., ..., Cao, B. (2020). Clinical features of patients infected with 2019 novel coronavirus in Wuhan, China. *The Lancet*, 395 (10223), 497–506.

Jia, P., Liu, L., Xie, X., Yuan, C., Chen, H., Guo, B., Zhou, J., Yang, S. (2021). Changes in dietary patterns among youths in China during COVID-19 epidemic: The COVID-19 impact on lifestyle change survey (COINLICS). *Appetite*, 158 .

Karlsson, L. K., Jakobsen, L. H., Hollensberg, L., Ryg, J., Midttun, M., Frederiksen, H., ..., Lund, C. M. (2021). Clinical presentation and mortality in hospitalized patients aged 80+ years with COVID-19 – a retrospective cohort study. *Archives of Gerontology and Geriatrics*, 94, 104335.

Khan, K. U., Zuberi, A., Batista Kochenborger Fernandes, J., Ullah, I., Sarwar H. (2017). An overview of the ongoing insights in selenium research and its role in fish nutrition and fish health. *Fish Physiol Biochemistry*, 43(6), 1689–1705.

Lewis, E. D., Meydani, S. N., Wu, D. (2019). Regulatory role of vitamin E in the immune system and inflammation. *IUBMB Life*, 71(4), 487–494.

Li, R., Wu, K., Li, Y., Liang, X., Tse, W. K., Yang, L., Lai, K. P. (2020). Revealing the targets and mechanisms of vitamin A in the treatment of COVID-19. *Aging* (Albany NY), 12(15), 15784.

Mikkelsen, K., Stojanovska, L., Prakash, M., Apostolopoulos, V. (2017). The effects of vitamin B on the immune/cytokine network and their involvement in depression. *Maturitas*, 96, 58–71. doi:10.1016/j.maturitas.2016.11.012

Moise, A. R., Noy, N., Palczewski, K., Blaner, W. S. (2007). Delivery of retinoid-based therapies to target tissues. *Biochemistry*, 46(15), 4449–4458.

Mora, J. R., Iwata, M., Von Andrian, U. H. (2008). Vitamin effects on the immune system: vitamins A and D take centre stage. *Nature Reviews Immunology*, 8(9), 685–698.

Munasir, Z., Akib, A., Beeler, J., Audet, S., Semba, R. D., Sommer, A. (1995). Reduced seroconversion to measles in infants given vitamin A with measles vaccination. *The Lancet*, 345(8961), 1330–1332.

Muscogiuri, G., Barrea, L., Savastano, S., Colao, A. (2020). Nutritional recommendations for CoVID-19 quarantine. *European Journal of Clinical Nutrition*, 74, 850–851.

Ng, T. B., Cheung, R. C. F., Wong, J. H., Wang, Y., Ip, D. T. M., Wan, D. C. C., Xia, J. (2015). Antiviral activities of whey proteins. *Applied Microbiology and Biotechnology*, 99, 6997–7008.

Opichka, K., Smith, C., Levine, A. S. (2019). Problematic eating behaviors are more prevalent in african american women who are overweight or obese than african american women who are lean or normal weight. *Family and Community Health*, 42, 81–89.

Pauling, L. (1970). Evolution and the need for ascorbic acid. *Proceedings of the National Academy of Sciences*, 67(4), 1643–1648.

Pauling, L. (1971). Vitamin C and common cold. *JAMA*, 216(2), 332–332.

Peterhans, E. (1997). Oxidants and antioxidants in viral diseases: Disease mechanisms and metabolic regulation. *Journal of Nutrition*, 127, 962S–965S.

Rahman M. T., Idid, S. Z.(2020). Can Zn Be a Critical Element in COVID-19 Treatment? *Biological Trace Element Research*, 1–9.

Richardson, D. P., Lovegrove, J. A. (2021). Nutritional status of micronutrients as a possible and modifiable risk factor for COVID-19: A UK perspective. *British Journal of Nutrition*, 125(6), 678–684.

Rodriguez-Besteiro, S., Tornero-Aguilera, J. F., Fernández-Lucas, J., Clemente-Suárez, V. J. (2021). Gender differences in the COVID-19 pandemic risk perception, psychology, and behaviors of Spanish university students. *International Journal of Environmental Research and Public Health*, 18(8), 3908.

Ruiz-Roso, M. B., de Carvalho Padilha, P., Mantilla-Escalante, D. C., Ulloa, N., Brun, P., Acevedo-Correa, D., Arantes Ferreira Peres, W., Martorell, M., Aires, M. T., de Oliveira Cardoso, L. (2020). Covid-19 confinement and changes of adolescent's dietary trends in Italy, Spain, Chile, Colombia and Brazil. *Nutrients*, 12(6), 1807.

Rundle, A. G., Park, Y., Herbstman, J. B., Kinsey, E. W., Wang, Y. C. (2020). COVID-19-related school closings and risk of weight gain among children. *Obesity*, 28, 1008–1009.

Semba, R. D. (1999). Vitamin A and immunity to viral, bacterial and protozoan infections. *Proceedings of the Nutrition Society*, 58(3), 719–727.

Semba, R. D., Akib, A., Beeler, J., Munasir, Z., Permaesih, D., Martuti, S. (1997). Effect of vitamin A supplementation on measles vaccination in nine-month-old infants. *Public Health*, 111(4), 245–247.

Sidor, A., Rzymski, P. (2020). Dietary choices and habits during COVID-19 lockdown: Experience from Poland. *Nutrients*, 12, 1657.

Song, P., Li, W., Xie, J., Hou, Y., You, C. (2020). Cytokine storm induced by SARS-CoV-2. *Clinica Chimica Acta*, 509, 280–287.

Tayyib, N. A., Ramaiah, P., Alsolami, F. J., Alshmemri, M. S. (2020). Immunomodulatory effects of zinc as a supportive strategies for COVID-19. *Journal of Pharmaceutical Research International*, 14–22.

Te Velthuis, A. J., van den Worm, S. H., Sims, A. C., Baric, R. S., Snijder, E. J., van Hemert, M. J. (2010). Zn2+ inhibits coronavirus and arterivirus RNA polymerase activity in vitro and zinc ionophores block the replication of these viruses in cell culture. *PLoS Pathogens*, 6(11), e1001176.

Thomas, W. R., Holt, P. G. (1978). Vitamin C and immunity: An assessment of the evidence. *Clinical & Experimental Immunology*, 32, 370–379.

Toyosawa, T., Suzuki M., Kodama, K., Araki, S. (2004). Highly purified vitamin B2 presents a promising therapeutic strategy for sepsis and septic shock. Infection and Immunity Mar; 72(3), 1820–1823.

Traber, M. G. (2007). Vitamin E regulatory mechanisms. *Annual Review of Nutrition*, 27, 347–362. doi:10.1146/annurev.nutr.27.061406.093819

Traber, M. G., Sies, H. (1996). Vitamin E in humans: Demand and delivery. *Annual Review of Nutrition*, 16(1), 321–347.

Webb, A. L., Villamor, E. (2007). Update: Effects of antioxidant and non-antioxidant vitamin supplementation on immune function. *Nutrition Reviews*, 65, 181–217.

Yoshii, K., Hosomi, K., Sawane, K., Kunisawa, J. (2019). Metabolism of dietary and microbial vitamin B family in the regulation of host immunity. *Frontiers in Nutrition*, 6.

Zhang, J., Taylor, E. W., Bennett, K., Saad, R., Rayman, M. P. (2020). Association between regional selenium status and reported outcome of COVID-19 cases in China. *American Journal of Clinical Nutrition*, 111(6), 1297–1299.

Ziegler-Heitbrock, H. W. L., Sternsdorf, T., Liese, J., Belohradsky, B., Weber, C., Wedel, A., Schreck, R., Bauerle, P., Strobel, M. (1993). Pyrrolidine dithiocarbamate inhibits NF-κB mobilization and TNF production in human monocytes. *Journal of Immunology*, 151, 6986–6993.

# 23 Life Post COVID-19 Pandemic
## *Opportunities and Challenges for Decision Makers*

*Raees Ahmad Khan, Alka Agrawal, and*
*Md Tarique Jamal Ansari*
Department of Information Technology, Babasaheb Bhimrao Ambedkar
University, Lucknow, Uttar Pradesh, India

## CONTENTS

## 23.1 INTRODUCTION

The COVID-19 pandemic has impacted communities and economies worldwide and threatens to reorganize our environment. While the effects of the crisis both enhance familiar risks and create new risks, changes of this scale often open up new opportunities for management of structural problems and for better construction. Although there has been a global pandemic risk for centuries, COVID-19 has been a problem for most people, health systems, societies, and governments all over the world. In the middle of unique challenges, uncertainties, and innumerable personal losses, policymakers are forced to take decisions on the management of the pandemic's immediate effect and its aftermath, decisions that influence the state of the nation for generations to follow. Remarkable approaches have been implemented by decision makers, the private sector as well as citizens in response to the COVID-19 pandemic that are likely to have a significant impact on our global economic environment (Lee et al., 2020; Ansari & Khan, 2021). What is assured is that during this worldwide crisis people and decision makers understand important lessons, and day-to-day life after COVID will definitely be changed completely. Therefore, this is the time to look forward to anticipated changes in the post-COVID era. The world in which we live currently is not really the same as it was before (McKee & Stuckler, 2020). The chains of development have stopped, roads have been abandoned, and the world economy has crashed. We have already seen that after revolutions, epidemics, and economic collapse, there is a mutual response to all crises to revert to stability and also to the comfortable. COVID-19 has spread worldwide and had numerous issues on privacy, information protection, security, and legislation (Ansari et al., 2020; Kumar et al., 2020). These are obstacles that push businesses and organizations to protect, and to look ahead to, their digital engagement initiatives (Ansari & Pandey, 2018; Ansari et. al, 2018; Ansari et al., 2021).

As of 28 April 2023, according to Indian government records India has the second largest number of Covid cases in the world, behind the United States only. In March of 2020, India enforced one of the most extreme national lockdown in the country, trying to force citizens to stay at home and closing down companies and prompting a migrant evacuation by millions.

COVID-19 has brought to light the difficulty companies have in balancing productivity and sustainability in the production process. Restructuring productivity and resilience is not a straightforward process. For businesses, resilience can be translated into sustainable corporate development; for societies, resilience is both a prerequisite as well as a result of significant economic development that prioritises factors such as inclusion, equality, as well as enhanced quality of life. Enhanced resilience would result in substantial increased economic costs for the company. Even so, if another comparable problem occurs, the cost of not performing may also be high. Government policymakers can face significant tax burdens if businesses choose to move with their most productive supply to alternatives to maintain security of supply. After resettlement of factories, higher output costs are likely to be borne by customers and financial costs, such as employment losses, are borne by FDI-deprived economies. These impacts could further reduce consumption that, due to COVID-19, is already down (Anukoonwattaka & Mikic).

Not only have our economies and wellbeing but also our governments been challenged by the COVID-19 pandemic. With the world taking emergency steps to tackle the crisis, fears have started to arise that some nations are using the circumstance to scale back constitutional rights as well as human rights. Coronavirus also underlines and worsens systemic inequities – ranging from insufficient health services to social security inequalities, digital divisions, and unequal accessibility to education; from the deterioration of the infrastructure to racial inequality and women's abuse – which themselves constitute challenges to democracy. However, this emergency may also be a chance to move forward. Journalist, campaigners, and many others falsely promoting "fake news," intense internet-policing and increasing internet surveillance, temporary suspension of elections. Since the start of the crisis, a range of countries have used this tactic in their attempts to regulate information flow and limit freedom of speech and freedom of association.

The rest of this chapter is organized as follows: Section 23.2 discusses the new normal in society post COVID-19. Section 23.3 and describe the different opportunities and challenges faced by decision makers due to the pandemic. Finally, conclusions are given in Section 23.5.

## 23.2  UNDERSTANDING THE NEW NORMAL

Currently the pace and scale of activities triggered by Covid-19 is disorienting in a world where significant changes usually take place at a very slow rate. They came so quickly and were so impactful that it is difficult to understand how revolutionary some of them have been. Pandemics have a way of doing this and creating a new normal in societies. In several European countries, the worldwide influenza outbreak of 1918 helped to establish national health programmes. The 1929 Depression and the Second World War rapidly contributed to the framework for an advanced welfare state. Covid-19 sped up economic and social thinking transformation towards new standards (Dobson, 2020). Following are some of the changes we have seen in the new normal.

### 23.2.1  HEALTHIER LIFESTYLE

During the lockdown, societies were forced to get back to the basics. Restaurants and shopping centers were closed so people started to eat better and consider in more depth our consumer culture. Pollution was reduced and environmental damage declined and people realized that the health of the earth requires a sustainable lifestyle. People would offer priority to an easy but satisfying lifestyle, lower consumption, and increase savings as a safeguards against potential uncertainty of income. The earth's security and the next generation will be the guiding force for much of our lives.

### 23.2.2 Upgraded Healthcare System

The COVID-19 pandemic emphasized that the health system needs to be able to cope with rapid spikes in patients' numbers. Emergency services are being tackled with the development of greater capacity in healthcare facilities. Digital health innovations are on the upswing and patient care is changing such as the use of contactless temperature reading and telemedicine. Individuals have also become more mindful about the importance of hygiene, wearing masks and using hand sanitizer. Security of health services has become a greater priority.

### 23.2.3 Digital Education Sector

Educational institutions have had to update their teaching methods with intelligent technology in order to minimize the chances of infection. Schools have gone virtual and parents have had to take a greater role in day-to-day learning since children have been at home with online learning.

### 23.2.4 Family-Oriented Civilization

Families have had to learn to work collaboratively in these tough times and share in the burdens of housework and care of children and old people. Since forms of entertainment like theater and sports facilities have been closed people have had to stay home and spend more time together. This has resulted in a greater focus on familial bonds.

### 23.2.5 Work from Home Culture

Work from home, teleconferencing, and virtual meetings have become the new work culture due to the need for social distancing and other Covid safety measures. Individuals have adapted and now value job opportunities that allow them to cultivate close family relationships.

## 23.3 DECISION MAKING DURING THE COVID-19 PANDEMIC

COVID-19 has required leaders to make decisions and take risks never before encountered in relation to worker and consumer protection, corporate strategy, and other vital matters. An improved framework for decision making will help render decisions more strategic and practical decisions during the prolonged instability of the pandemic. A 2020 Gartner CEO Survey demonstrated that only 38% of CEOs rely on consistent decision-making processes and values that direct decision-making and overcome conflicts between the executive team. The remaining respondents use socialization strategies, trusted experts, and delegation. COVID-19 has made decision making more difficult because of the extraordinary characteristics of the emergency. The decision making process containing three equally relevant sets of criteria must therefore be followed by the representatives (Gartner, 2020). Gartner's three-part executive decision-making framework includes three criteria in the time of COVID-19 pandemic, as can be seen in Figure 23.1.

- Traditional business value: To safeguard the economic stability of the company and its profitability.
- Crisis and disruption: To safeguard worker, consumer, and societal wellbeing and security.

**FIGURE 23.1**   Decision-making criteria during COVID-19.

- Social and emotional: To safeguard the community's mental wellbeing and incorporate the company value system in such a way that represents current social objectives.

Three sets of parameters are equally relevant in the decision-making context as they reflect the business tradeoffs made by all management during a pandemic, where decision-making in public and emergency situation play an increased workload. In regular business conditions, while weighting may seem less acceptable, managers could give equal weight for three sets of parameters in a longer-term disturbance such as a pandemic.

## 23.4   CHALLENGES AND OPPORTUNITIES FOR DECISION MAKERS

The outbreak of COVID-19 has created both challenges and opportunities for businesses. Some businesses have had to turn to new supply chains, new ideas, and new operational strategies without adequate time to carefully manage their effects. Many decision makers are thus starting from the beginning, without their normal networks, to communicate closely with stakeholders and to reach consensus on the way ahead. Other problems for businesses have been taken on by the quick transition to working remotely. The challenge is always successful performance assessment, management, and transparency. How do we identify and reward great success and face performance problems in this almost entirely virtual environment, because it is much easier for workers to become almost hidden. More than ever, decision makers must be adapted to their corporation and needs of the people, considering the potential rise in tension around their own lives during the COVID-19 pandemic.

Although the last few years have seen steady improvement in electronic conferencing as well as remote working arrangements, several sensitive discussions or relationships in companies are still common in person. The best way to start and retain accountability and confidence is to promote clear communication, expressing of information and respond on responsibilities. The crucial challenge for decision makers in the current pandemic situation is how senior executives can interact with interested parties and internal management team in strategic decision making in ways that will improve trust, openness, and collaboration. COVID-19 is also a crucial opportunity for a revision of the delivery and management of decision-making regionally and abroad.

## 23.5 CONCLUSIONS

The COVID-19 tragedy has had an impact on businesses and civilizations all across the world, and as the issue develops, this would fundamentally change our world. Transformation of this magnitude also generates new possibilities for addressing systemic difficulties and methods to build back better, increasing both existing dangers and developing new ones. A sustainable society is developing as people are working together to face the challenges of Covid. People understand the need for precautions and the need for actions to combat future eventualities in order to safeguard the generations to come. There seems to be a higher risk chance that the global community could be even more separated, contentious, and nationalism after COVID-19. But rather an alternative interpretation is within striking range. Decision-makers need to inspire people to take advantage of the chances this crisis presents to enhance the state of the world by providing a thorough picture of predicted long-term developments.

## REFERENCES

Ansari, M. T. J., & Khan, N. A. (2021). Worldwide COVID-19 vaccines sentiment analysis through Twitter Content. *Electronic Journal of General Medicine, 18*(6).

Ansari, M. T. J., & Pandey, D. (2018). Risks, security, and privacy for HIV/AIDS data: Big data perspective. In *Big Data Analytics in HIV/AIDS Research* (pp. 117–139). IGI Global.

Ansari, M. T. J., Al-Zahrani, F. A., Pandey, D., & Agrawal, A. (2020). A fuzzy TOPSIS based analysis toward selection of effective security requirements engineering approach for trustworthy healthcare software development. *BMC Medical Informatics and Decision Making, 20*(1), 1–13.

Ansari, M. T. J., Baz, A., Alhakami, H., Alhakami, W., Kumar, R., & Khan, R. A. (2021). P-STORE: Extension of STORE methodology to elicit privacy requirements. *Arabian Journal for Science and Engineering*, 1–24.

Ansari, M. T. J., Pandey, D., & Alenezi, M. (2018). Store: Security threat oriented requirements engineering methodology. *Journal of King Saud University-Computer and Information Sciences*.

Dobson, J. (2020, May 23). Covid-19 is accelerating the New Normal. Retrieved October 11, 2020, from www. Sunday guardianlive.com/news/covid-19-accelerating-new-normal

Executive Decision Making Framework in the Time of COVID-19. (n.d.). Retrieved October 14, 2020, from www.gartner. com/en/insights/framework-for-executive-decision-making-in-the-time-of-covid-19

Kumar, R., Alenezi, M., Ansari, M. T. J., Gupta, B. K., Agrawal, A., & Khan, R. A. (2020). Evaluating the impact of malware analysis techniques for securing web applications through a decision-making framework under fuzzy environment. *International Journal of Intelligent Systems, 13*(6), 94–109.

Lee, S., Hwang, C., & Moon, M. J. (2020). Policy learning and crisis policy-making: Quadruple-loop learning and COVID-19 responses in South Korea. *Policy and Society, 39*(3), 363–381.

McKee, M., & Stuckler, D. (2020). If the world fails to protect the economy, COVID-19 will damage health not just now but also in the future. *Nature Medicine, 26*(5), 640–642.

# 24 Post COVID-19 Paradigm Shift
## A New Era of Digital Experience

*Alka Agrawal, Md Tarique Jamal Ansari, and
Raees Ahmad Khan*

Department of Information Technology, Babasaheb Bhimrao
Ambedkar University, Lucknow, India

## CONTENTS

## 24.1 INTRODUCTION

COVID-19 is a global pandemic with high communicability and collaborative initiatives by governments around the world have been focused on containment and prevention. Countries with low per-capita death rates of COVID-19 seem to share strategies that include early monitoring, testing, contact tracking, and strict quarantine. Collaboration and data processing in many countries depend on digital technologies being incorporated and integrated into policies and health care (Whitelaw et al., 2020). Millions of people across the world are involved in a mass workload initiative, culminating in unique criteria for executives to direct and inspire their workers. In such a context, more than almost any other disturbance (such as technological ones) in past years, the Covid- 19 pandemic has transformed the way we operate, transact, and socialize with people. Worldwide lockdowns have boosted digital technology adoption. In order to cope with such circumstances, companies, educational organizations, analytics, computers, data processing approaches, and online education technologies have been compelled to work together to boost production and performance time. A crucial element of coping with Covid- 19 is ensuring that healthcare services are provided to any extent necessary. Obviously, the pandemic has prompted everybody to focus on digital and interactive ways of working in the world. Because of Covid- 19, the government's Digital India dream is emerging as a critical resource for overcoming the present problem. People are realizing virtual ways of doing new things amid lockdowns that they thought were impossible before such

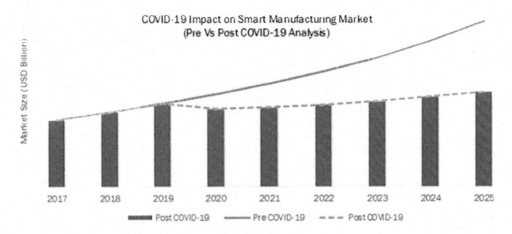

**FIGURE 24.1** COVID-19 impact on smart manufacturing market by enabling technology, information technology, industry, and region – global forecast to 2025.

*Source*: Press release, Investor relation presentation, annual report, expert interview, and markets analysis.

as virtual schooling, teleconferencing, and mobile applications for doing routine tasks. As a result of the COVID-19 pandemic, organizations have revamped the workplace and introduced new corporate policies to align with social distancing and other public health guidelines. Digital transformation has taken steps to fill the gaps of enforced shutdowns as well as social distancing. Large-scale emerging developments include the Internet of Things (IoT), 5G telecommunication networks of the next decade, large-scale data mining, and artificial intelligence (AI) that uses machine learning and blockchain technology (Ting et al., 2020; Zarour et al., 2020). COVID-19 has stressed the need for effective new platforms for pandemic planning, monitoring, research, tracing, quarantine, and incorporating them into their systems (Whitelaw, 2020).

Post COVID-19, the worldwide market in intelligent smart manufacturing is estimated to rise from USD 181.3 billion in 2020 to USD 220.4 billion by 2025, at 4.0% CAGR. The growing demand for intelligent products and solutions driven by COVID-19 have enhanced the digital manufacturing market (M&M). Figure 24.1 shows the study of Markets and Markets demonstrating COVID-19 impact on smart manufacturing market by enabling technology, information technology, industry, and region.

Different sectors have influenced the human experience and resulted in significant motivation in digital services, as will be discussed. The rest of this study has been organized as follows: Section 24.2 discusses the COVID-19 pandemic and digital technology innovations in India. Section 24.3 describes the significant impact of COVID-19 on different sectors. Several opportunities and challenges for digital technologies due to the pandemic are discussed in Section 24.4. Section 24.5 discusses the results of this study. Finally, the conclusions are given in Section 24.6.

## 24.2   COVID-19 AND DIGITAL TECHNOLOGY IN INDIA

Digital development in India was already under way when the pandemic impacted. However, COVID-19 pushed Indian decision-makers to concentrate on technology use in different fields. Recently, emerging technology innovations in key areas and segments such as healthcare, food safety, education, and the whole supply chain have led to key outcomes. Emerging technologies have put digital technologies at the frontline of efforts to tackle the COVID-19 pandemic in India by central as well as state governments. In the rush to implement such digital solutions, little introspection has been given as to whether legal and technological structures will support these systems to guarantee that public

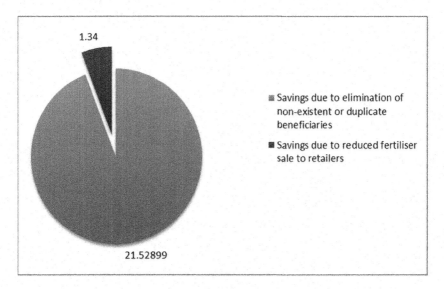

**FIGURE 24.2** Estimated savings/benefits (in USD billions).

health and social confidence are not impeded by them. The government of India called on corporate and market sectors to encourage "Atma Nirbhar Bharat," which means self-reliant India. A number of actions have been announced by the Government of India to implement major transformation in areas like agriculture, infrastructure, education, and healthcare, with the goal of renewing investor confidence and improving the government's Make in India Campaign. Currently, the private sector is planning to invest in advanced technology initiatives that will transform the economy, provide possibilities for employment, and inspire customer sentiment (Jalan et al., 2020).

Approximately 100 million digital transaction statistics are now being tracked every day with 5 trillion rupees ($67 trillion) of values as per the data from the apex Reserve Bank of India (RBI), which is a five times increase from 2016. The RBI expects this to further grow five-fold to 1.5 billion transactions a day worth 15 trillion rupees ($200 billion). Most of this comes from the United Payment Interface (UPI), a reliable payment system established by National Payments Corporation of India and controlled by RBI. In addition to technical development, this first digital reset of a nation of 1.3 billion people is the centerpiece of a new framework for policy delivery. Two historically difficult commitments in India – speed and plugging of leaks – are integrated in this conceptual use of state technology. The implementation of digital technologies has saved almost $23 billion, of which 98 percent has resulted in the removal of fake beneficiaries (Sharma, 2020). This report's graphical representation can be seen in Figure 24.2.

Indian states have taken every opportunity amid COVID-19 to further promulgate technical use, such as by spraying disinfectants by drones, tracking containment regions, and making public alerts. The use of technology decentralizes decision-making and bridges neighborhoods around towns and cities with local authorities. These creative technologies have an impact on different areas of life, whether livelihoods, infrastructure, or education, through technology and digital resources.

## 24.3 COVID-19 IMPACT ON DIFFERENT SECTORS

Covid has resulted in changes in how we communicate with friends and family, how we work and travel, manage our healthcare, spend our free time, and perform daily activities. These modifications

**FIGURE 24.3** Impact of COVID-19 on different sectors.

have driven the transition into digital technology in all sectors to an impressive degree and pace. Figure 24.3 shows the different sectors that have been affected by COVID-19 and the need for technological advancements.

### 24.3.1 AGRO & FOOD PROCESSING

Agriculture is known as the Indian economy's foundation. With the facilities of interstate transport closure, farmers cannot sell their crop production. They lose heavily and have to throw their crops out. They have no other revenue source. The poultry industry, which is Indian economy's quickest increasing subsector, has also suffered significant loss because of social media where misunderstanding has been circulated by corresponding COVID-19 contamination with meat and poultry production. While the government offers assistance, the circumstances are always miserable. Technology plays a significant role in farming and agricultural practices. Boosting farmers' livelihoods and agricultural yield is dependent on the adoption of advanced farming technology as well as the rate at which agriculture are digitized.

### 24.3.2 E-COMMERCE

The government released special advice on preserving social isolation in order to avoid the transmission of COVID-19 from the group. The domestic lockdown will have an enormous impact on e-commerce services, particularly at a time of booming market for home deliveries. Their losses could be compensated with flexible financial and service policy could be applied by the government such as loss-making enterprises to E-commerce and the permit to operate subject to certain restrictions would be given.

### 24.3.3 EDUCATION & LEARNING

The emergence of the COVID-19 pandemic led to the closing of schools to decrease disease transmission among students in schools and colleges. The education sector has been forced to move online. Many academic institutions with technical capacities offer live courses such as Byjus, Extra-marks, etc. The government was compelled to improve the edtech sector amid the COVID-19 pandemic. Long-term learning methods of COVID-19 have improved. Technology has been used to develop digital platforms to help students keep learning and ensure teachers stay employed.

### 24.3.4 SAFETY & SECURITY

COVID-19 influenced defence industries' supply chains and development processes. They must rely on various components from different sources in the countries concerned. The demand for defence equipment would also decrease. The current situation is not even suitable for business growth, because we know that during defence exhibitions, many high-value acquisition projects were finalized and cancelled. Military drills that expose the potential customer to international equipment and their expertise also have had a negative effect on business growth since many countries such as United States and the United Kingdom have rescinded travel plans, interventions, and military exercises. Even the mounted equipment that is ready for shipment is held due to the aviation lockdown. Because of non-delivery on time, their sales would decline dramatically, which will impact manufacturers' balance sheets in turn. COVID-19 has shown the defence industry that the various facets of risk preparation have to be explored. People must move to technological platforms or begin to use an automated system.

### 24.3.5 HOSPITALITY & TRAVEL

Towards a complete ban on domestic and international airlines, tourism revenues have dropped. Even a lot of visitors have forced to cancel them. Meetings, conventions, and big international activities such as international congresses, Olympics, Wimbledon, International Film Festival in Cannes, and Facebook F8 have been cancelled, leading to massive losses. Before that, many Indians travelled both domestically and globally, and now no one is prepared to go anywhere. The Indian Association of Tour Operators has suggested that, as a result of the Indian government's restrictions on flight transportation, the hotel, airline, and travel sector may lose 8,500 crores collectively.

### 24.3.6 HEALTHCARE SERVICES

The weaknesses in health systems have been revealed by COVID-19. While accessibility to health care is an important right fear of COVID-19 has influenced the primary healthcare arrangements of individuals. For the expectant mothers the pandemic stopped the obstetrician from attending for pre-natal exams, which made telemedicine difficult to perform. This pandemic has shown that temples, sculptures, and monuments are not important, but world-class hospitals are. The competitiveness of the producer of medical devices that imports consumables and products from China may also be adversely affected.

## 24.4 OPPERTUNITIES AND CHALLENGES FOR DIGITAL TECHNOLOGIES

During COVID-19, digital technology has become a great source of information for society. It has empowered societies to create unity and give mutual support and encouraged us to pursue social relations in the midst of pandemic and lock-up. Nevertheless, we need to be cautious against the misplaced dependency on technologies that exclude and threaten our approaches to the pandemic, but that adversely affect and benefit socialist principles. The following subsection describes the different opportunities and challenges for the adoption of digital technologies in the pandemic.

### 24.4.1 OPPORTUNITIES

(i)   Digital transformation: More and more technology transformation, ecosystem proposals, and emerging technologies are making up most of the economic value generated. The high appreciation of major technology and quick investment in this room is clearly evident. This can be seen here. In these times of difficulty, however, countries can try to maintain their advantage and raise non-tariff barriers to pioneering digital research and development, intellectual property, and goods or solutions they perceive to be essential to domestic safety or the economy. Enhanced regulatory structures and enhanced inspection of M&A activities can be expected.

(ii) Digital culture: Although electronic development and work has already been done in the world, COVID-19 should speed up the step towards greater automation. Companies would like to use digital channels progressively that can proceed in case of COVID-19 emergency and reduce employee reliance. In addition, the marginal pressure will stimulate companies to aim for productivity in the implementation of digital resources. In reality, this can be the optimal opportunity to use these digital technologies in general.

(iii) Digital workspaces: COVID-19 may be a breakthrough in the way individuals work. The pandemic affects the people's attitude to jobs, mobility, and versatile models of jobs. We believe that alternative model implementation of work and creation of jobs will be accelerated. Digital networks for revelation will supply many of these emerging models.

(iv) Digital service models: Consumers/companies may aim to include manufacturers or collaborators in the outputs and not in conventional content and supplies with more market instability. We anticipate several companies to reconfigure their proposals and to connect their solutions with the consequences/service models into this new environment. Either online platform may form the foundation of such proposals.

(v) Digital skills: Digital technology-driven advancement may gradually boast companies' competitive advantage on the market. Triad of abilities – field/functional knowledge, technical skills, and specialized skills – would develop the positions of the coming years. This would be a good time for companies to think about the virtual talent pool and use their downtimes to create an all-round base for digital skills.

(vi) Tech foundations: Technological debt accumulates over time in several big corporations as companies experience different technology implementation cycles. This crisis could provide leaders with the chance to examine the technological environment closely and find ways to withdraw this technological debt. It can also produce valuable cash flows that can be used for potential digital development.

### 24.4.2 CHALLENGES

(i) Legislative requirement: There is a need for effective process at the moment to track the use of monitoring or decision-making technology, especially if deployed by the community or even in public organizations. Legal mechanisms need to emphasize accountability by specifying what information is being gathered about people and by creating a constitutional requirement to use this data only within the system of public wellbeing, which can protect against its functioning. The legislation, ultimately, shall set the timing of these measures by continuing parliamentary supervision or by a negotiated deal ensuring that the initiatives for monitoring do not proceed after the pandemic.

(ii) Socio-economic inequality: Digital transformation programs can intensify socio-economic inequality and lead to differences in healthcare prevalence. The use of the internet and digital technology is generally part of modern technology. While the worldwide use of the Internet in 4 billion users in 2019 was disproportionately higher in high-income areas than those in low-income and middle-income nations. Susceptible communities, such as low-income communities or remote areas, really are not able to gain internet access even in high-income countries. Initiatives should be customized to the target regions to successfully introduce digital technologies globally (Wallis et al., 2020).

(iii) Security and privacy concern: There would be a large increase in cybersecurity threats in the COVID-19 situation. Worldwide, ransomware, fake COVID-19 applications, and automated phishing scams have shown a strong rise. Due to COVID-19 online transactions and interaction rely progressively on digital applications through the adoption of electronic networks and software, application security and data privacy has become more and more required (Ansari et al., 2018; Kumar et al., 2020). Elicitation of efficient software security requirements needs a clear understanding of security standards and their relevance in the specification of application. (Ansari et al., 2020, 2021).

(iv) Digital surveillance: The negative side of digital workplace and concert work can be supposed to boost concerns of tension, appearance, job overload, tracking and surveillance. It is important to consider and analyze new and extreme ways of digital surveillance or monitoring of digital workplace (Rahul et al., 2020).

## 24.5   DISCUSSION

Digital technology has increased the ability of individuals to work from home (WFH), but remote work is subject to substantial variations all over countries, depending on the nature of the job and activities to be performed and digital skills. Employment favorable to working remotely in developed nations, higher-level training employees and wage-earning full-time jobs are widespread. The chances of working remotely are considerably lower for women and young employees. In several developing nations, digital technology is limited or of poor quality. The rapid proliferation of pandemic-accelerated technologies has contributed to an immediate need to adapt companies and governments. Most companies are technologically separated, particularly those in developing countries. They do not have access to appropriate workforce and face demanding market conditions. On the other hand, workers have really no security and are unable to adjust to the labour market with skill or versatility. In order to meet these obstacles, organizations should implement technologies and redesign training programmes to give their employees technical potential. Companies should also consider training to allow potential employees to have the required skill set. Governments are given an abundance of policy choices, from rewards and laws to technology and development initiatives.

Some key priorities that should be considered by government/organizations are:

- Investment in human resources and higher education should be stepped up to make significant changes workforce to future job opportunities.
- Enhance social safety, extend security network coverage, and change funding systems and labour force requirements to promote transfers and decrease regulatory barriers to formal employment.
- Ensure reliable internet access and adjust legislation to overcome electronic platforms' problems such as data security and competition legislation.
- Improve transfer pricing mechanisms and create financial resources for fundamental social security and growth of human resources.

Public health will become more and more digital and thus it is imperative to emphasize the implications of digital technologies. Various stakeholders, such as technology businesses, must be long-term readiness collaborators rather than partners only in current pandemic situations. COVID-19 does not respect any government systems, cultural conventions, or geographical or social boundaries and emerging technology in addition data are progressively unknown. Global approaches to monitor, assess, and employ emerging technologies are urgently required in order to improve pandemic control and strategic planning for COVID-19 as well as other serious diseases (Budd et al., 2020). Technology could be a benefit to social systems if corporations and administrations plan and make adjustments. The COVID-19 pandemic has taken communities to an observation point whereby technology is not a choice but a requirement. Corporation and policymakers can use this pandemic situation to look forward to the future with the right measures and procedures. The COVID-19 pandemic has demonstrated to us that new technologies, such as the Internet of Things and artificial intelligence, are not just resources for our culture and our economies. We really have to conceive of them as crucial infrastructure, especially in this period of adjustment. It has never been so important to change, concentrate on people, and be responsive in the way we implement new technology policies and protocols. If we could somehow adapt technology to the required structures, standards, and guidelines in reaction to this COVID-19 pandemic, we would become bigger and more powerful than ever.

## 24.6  CONCLUSIONS

COVID-19 has forced people all over the world to reassess their operational practices. This situation demands more, rather than less, sustainable digital development. Technological innovation can be powerful leverage for companies to adapt to the present situation and provide an opportunity to reconfigure their business and designs for the future. This might be appropriate to re-consider digital projects based on importance in current circumstances. Across various industries, including manufacturing, financial services, medical care, and education, adoption of technology would be more effective in encouraging their performance levels and guaranteeing the sustainability of expertise that promotes smooth functioning in a workplace environment. There can be more pressing new challenges and opportunities. Any electronic strategy for innovation that does not produce any benefit at any rate must be rethought. The objective is to continue experimenting and innovating, taking advantage of the opportunities the pandemic has presented.

## REFERENCES

Ansari, M. T. J., & Pandey, D. (2018). Risks, security, and privacy for HIV/AIDS data: Big data perspective. In *Big Data Analytics in HIV/AIDS Research* (pp. 117–139). IGI Global.

Ansari, M. T. J., Al-Zahrani, F. A., Pandey, D., & Agrawal, A. (2020). A fuzzy TOPSIS based analysis toward selection of effective security requirements engineering approach for trustworthy healthcare software development. *BMC Medical Inormatics and Decision Making, 20*(1), 1–13.

Ansari, M. T. J., Baz, A., Alhakami, H., Alhakami, W., Kumar, R., & Khan, R. A. (2021). P-STORE: Extension of STORE methodology to elicit privacy requirements. *Arabian Journal for Science and Engineering,* 1–24.

Ansari, M. T. J., Pandey, D., & Alenezi, M. (2018). STORE: Security threat oriented requirements engineering methodology. *Journal of King Saud University-Computer and Information Sciences.*

Budd, J., Miller, B. S., Manning, E. M., Lampos, V., Zhuang, M., Edelstein, M., ... & Short, M. J. (2020). Digital technologies in the public-health response to COVID-19. *Nature Medicine,* 1–10.

Jalan, M., Chawla, S., & Sudhakaran, S. (2020, June 30). Mapping Digital Transformation in India During the COVID-19 Pandemic. Retrieved October 07, 2020, from https://apcoworldwide.com/ blog/ mapping-digital-transformation-in-india-during-the-covid-19-pandemic

Kumar, R., Alenezi, M., Ansari, M. T. J., Gupta, B. K., Agrawal, A., & Khan, R. A. (2020). Evaluating the impact of malware analysis techniques for securing Web applications through a decision-making framework under fuzzy environment. *International Journal of Intelligent Systems, 13*(6), 94–109.

Rahul De, N. P., & Pal, A. Impact of digital surge during covid-19 pandemic: A viewpoint on research and practice. *International Journal of Information Management.*

Ting, D. S. W., Carin, L., Dzau, V., & Wong, T. Y. (2020). Digital technology and COVID-19. *Nature Medicine, 26*(4), 459–461.

Wallis, L., Blessing, P., Dalwai, M., & Shin, S. D. (2017). Integrating mHealth at point of care in low-and middle-income settings: The system perspective. *Global Health Action, 10*(Suppl 3), 1327686.

Whitelaw, S., Mamas, M. A., Topol, E., & Van Spall, H. G. (2020). Applications of digital technology in COVID-19 pandemic planning and response. *The Lancet Digital Health.*

Written by Ankita Sharma, S. (n.d.). COVID-19 has accelerated India's digital reset. Retrieved October 06, 2020, from www.weforum.org/agenda/2020/08/covid-19-has-accelerated-india-s-digital-reset/

Zarour, M., Ansari, M. T. J., Alenezi, M., Sarkar, A. K., Faizan, M., Agrawal, A., ... & Khan, R. A. (2020). Evaluating the impact of blockchain models for secure and trustworthy electronic healthcare records. *IEEE Access, 8,* 157959–157973.

# 25 COVID-19

## *Our Solutions Are in Nature*

Aparna Sarin

Senior Scientific Officer, Uttarakhand State Council for Science &
Technology, Dehradun, Uttarakhand, India

## CONTENTS

> As we encroach on nature and deplete vital habitats, increasing numbers of species are at risk.
> That includes humanity and the future we want.
>
> *UN Secretary-General Antonio Guterres*

## 25.1 INTRODUCTION

The COVID-19 pandemic is the defining global health crisis of our time and the greatest challenge we have faced since World War II. Since its emergence in Asia in late 2019, the virus has spread to every continent. The sudden outbreak of COVID-19 happened in December 2019 in Wuhan City of Hubei Province in China and spread to other parts of China and overseas. The World Health Organization declared a Public Health Emergency of International Concern regarding COVID-19 on 30 January 2020, and later declared a pandemic on 11 March 2020. On February 7, Chinese researchers said the virus could have spread from an infected animal to humans through illegally trafficked pangolins, prized in Asia for food and medicine. Scientists have pointed to either bats or snakes as possible sources. With a global death toll of more than 3 million and more than 145.4 million confirmed cases (https://covid19.who.int/), COVID-19 has emerged as one of the deadliest pandemics in history. The United States, India, and Brazil have reported the highest number of cases followed by France, Russia, and the UK.

The coronavirus family causes illnesses ranging from the common cold to more severe diseases such as severe acute respiratory syndrome (SARS) and Middle East respiratory syndrome (MERS), according to the WHO. Zoonotic infections have always featured among the wide range of human diseases and most (e.g., anthrax, tuberculosis, plague, yellow fever, and influenza) have come from domestic animals, poultry, and livestock (Wang and Crameri, 2014). However, with changes in the environment, habitat, and human behaviour, increasingly these infections are emerging from wildlife species to the extent that wildlife is thought to be the source of at least 70% of all emerging diseases (Kuiken et al., 2005; Wang and Crameri, 2014). For example, the expansion of road networks and the development of agricultural land have facilitated the spread of Nipah virus, West Nile virus, influenza A H5N1, monkeypox, SARS, HIV, and other novel pathogens throughout the world (Daszak, 2012; Kilpatrick and Randolph, 2012; Jones et al., 2013; Johnson et al., 2015).

Anthropogenic changes such as deforestation, habitat fragmentation, land-use, agricultural development, and uncontrolled urbanization can impact the transmission of infectious disease from animals to humans by altering the biodiversity's dilution effect. These changes also alter pathogens' niches, and the movements of hosts and vectors preparing the general ground that favors the emergence of infectious diseases.

## 25.2   WHY IS BIODIVERSITY IMPORTANT?

Biodiversity, or biological diversity, is the multitude of living things that make up life on Earth. It encompasses the 8 million or so species on the planet, from plants and animals to fungi and bacteria, and the ecosystems that house them such as oceans, forests, mountain environments, and coral reefs. On land, the most important ecosystems and biodiversity refuges are forests, which are home to most of Earths terrestrial biodiversity. Biodiversity plays several significant roles in the following ways.

### 25.2.1   Food Security

Biodiversity directly contributes to food security, nutrition, and well-being through a variety of plant and animal sources, both domesticated as well as wild. Human health and productive livelihoods depend upon ecosystem products and services such as fresh water, food, and fuel sources. Today, 75% of the worlds food is generated from only 12 plants and five animal species (WHO, 2019). In the last century, about 75% of plant genetic diversity has been lost as farmers worldwide have left cultivation of various local varieties and landraces in favour of mono-culturing genetically uniform, high-yielding varieties. As a result, out of the 250,000 to 300,000 known edible plant species, only 4% (i.e., 150–200) are used by humans. Animals provide some 30% of human requirements for food and agriculture and 12% of the worlds population lives almost entirely on products from ruminants.

### 25.2.2   Nutritional Impact of Biodiversity

Access to a sufficient and nutritious variety of food is a fundamental determinant of human health. Nutrition and biodiversity are interlinked with the species in the ecosystem and the genetic diversity within species. Genetically diverse resources of crops, livestock, and marine species ensure the sustainable productivity of soils and influence world food production. Availability of micronutrients in the diet can differ dramatically between different foods as well as among varieties/cultivars of the same food. Healthy local diets, with adequate average levels of nutrients intake, play a significant role in ensuring maintenance of high biodiversity levels.

Intensified and enhanced food production through irrigation, use of fertilizer, plant protection (pesticides), or introduction of specific crop varieties and cropping patterns affect native biodiversity, nutritional status, and human health. Habitat simplification and species loss often make communities vulnerable to poor health.

### 25.2.3   Importance of Biodiversity for Health Research and Traditional Medicine

For millennia, people all over the world have been using traditional medicine, herbal, or animal-derived remedies in health care, especially in primary health care. In some countries, traditional medicines using knowledge handed down through generations are extensively incorporated into the public healthcare system.

In Africa, Asia, Latin America, and the Middle East, 70–95% of the population still uses traditional medicine for primary health care. And some 100 million people are believed to use traditional, complementary, or herbal medicine in the European Union (EU) alone, as high as 90% of the population in some countries (Robinson and Zhang, 2011).

Medicinal plants are the most widely used medication tool in traditional and complementary medicine. Medicinal plants are collected from wild populations and also specially cultivated (WHO, 2015). Although synthetic medicines are available for many purposes, the global need and demand for natural products persists for use as medicinal products and also for biomedical research involving plants, animals, and microbes to understand human physiology and treatment of human diseases.

### 25.2.4  INFECTIOUS DISEASES

Anthropogenic activities are altering native biodiversity by disturbing both the structure and functions of ecosystems. Such disturbances are altering the interactions between organisms and their physical and chemical environments. Patterns of infectious diseases are sensitive to these disturbances. Landscapes and habitats have a variety of niches supporting a number of vertebrate and invertebrate species, and each species or taxon supports an even more impressive array of macro- and micro-parasites. Pathogens originating in wild animals have become increasingly important throughout the world recently because of the substantial impacts on human health, agricultural production, wildlife-based economies, and wildlife conservation. The exponential rise of human activity is one of the major reasons of health issues arising out of close encounters with such wild animals. The ever-increasing human population, increased migration, and international travel and trade, deforestation, land-use change, uncontrolled urbanization, movement of animals and animal products, and a range of environmental changes favour the transfer of pathogens between wild hosts, vectors, and domestic animals.

### 25.2.5  CLIMATE CHANGE, BIODIVERSITY, AND HEALTH

Climatic conditions significantly influence the biodiversity of various ecosystems. Hence, human health is impacted directly and indirectly by changing climatic conditions. While terrestrial biodiversity is influenced by extreme weather events (e.g., drought, flooding) marine biodiversity is affected by ocean acidification. Long-term changes in climate affect the viability and health of ecosystems, influencing shifts in the distribution of plants, pathogens, animals, and even human settlements.

### 25.2.6  THREATS TO BIODIVERSITY AND HEALTH

Due to anthropogenic pressure, species are being lost at a rate 1,000 times faster than at any other time in recorded human history. The world is facing an extinction crisis that is unprecedented in both magnitude and pace (Ostfeld, 2009). One million plant and animal species are threatened with extinction due to changes in land and sea-use, overexploitation, climate change, pollution, and invasive alien species (Diaz et al., 2019). Populations of mammals, birds, reptiles, amphibians, and fish have declined on average by 68% since the 1970s and vast areas of ecosystems have been degraded (WWF, 2020). At current rates, global extinctions within some classes of vertebrates are predicted to approach 50% within about 100 years (Eldredge, 2002). Extinction of local populations and metapopulations is also rampant but is poorly documented. The World Economic Forum (WEF) has recently noted that the world has lost 60% of all wildlife in the last 50 years while the number of new infectious diseases has quadrupled in the last 60 years. Already, the roughly 1-degree Celsius rise in mean global temperatures is affecting the abundance, genetic composition, behaviour, and survival of some species.

There is growing concern about the health consequences of biodiversity loss. Land-use change is the leading driver for emerging zoonoses and is likely to increase in the future (Loh et al., 2015). Biodiversity changes affect ecosystem functioning and significant disruptions of ecosystems can result in life sustaining ecosystem goods and services. Biodiversity loss also means that we are losing, before discovery, many of natures chemicals and genes, of the kind that have already provided humankind with enormous health benefits. Loss in biodiversity may limit discovery of potential treatments for many diseases and health problems.

## 25.3 CONNECTION BETWEEN ZOONOTIC DISEASES AND BIODIVERSITY

Natural environments are being destroyed by humankind at an accelerating rate. Between 1980 and 2000, more than 100 million hectares of tropical forest were felled, and more than 85% of wetlands destroyed since the start of the industrial era. In so doing, human populations, often in precarious health, have been put in close contact with new pathogens. The disease reservoirs are wild animals usually restricted to environments in which humans are almost entirely absent or who live in small, isolated populations.

### 25.3.1 ZOONOTIC DISEASES

Zoonotic diseases cause millions of deaths every year (Can et al., 2019). Zoonoses are the diseases transmitted from other animal species to human beings and account for approximately 60% of all infectious diseases and 75% of emerging infectious diseases in humans (Taylor et al., 2001). The primary drivers of zoonotic diseases are changes in the environment, usually the result of human development or climate change. Due to the destruction of forests, villagers settled on the edge of deforested zones hunt wild animals and send infected meat to cities, this is how Ebola found its way to major human centres. Exotic and wild species have been hunted to the brink of extinction for purely recreational reasons because of the physical appeal of rare species, exotic meals, naive pharmacopeia, etc. The trade in rare animals feeds the markets and in turn leads to the contamination of urban centres by new maladies. A recent study found that in areas under significant human use (e.g., agricultural and urban systems), wildlife hosts of human pathogens account for a greater share of total species abundance (21–144% higher) and species richness (18–72% higher) than in nearby undisturbed ecosystems (Gibb et al., 2020).

In addition to COVID-19, examples of emerging zoonotic diseases that have caused human health crises include Ebola, avian influenza, sudden acute respiratory syndrome (SARS), Middle East respiratory syndrome (MERS), and Human Immunodeficiency Virus (HIV). Evidence for an association between disease emergence and ecotones has been documented for several zoonoses such as Nipah virus encephalitis, influenza, rabies, and hantavirus pulmonary syndrome. The Ebola virus outbreak in West Africa was the result of deforestation, leading to closer contact between humans and wildlife. The avian flu was linked to intensive poultry farming and the Nipah virus resulted from the intensification of pig farming in Malaysia. The epidemic of severe acute respiratory syndrome (SARS) rose out of the proximity between bats, carnivores, and gullible human consumers. The U.S. Centre for Disease Control and Prevention estimates that three-fourths of new or emerging diseases that infect humans are zoonotic diseases.

The World Health Organisation (WHO) and most infectious disease experts agree that the origins of future human pandemics are likely to be zoonotic, with wildlife emerging as the primary source (Wang and Crameri, 2014). Illegal wildlife trade, the fourth-largest global crime is another important driver of infectious diseases and represents a serious threat to global health due to its clandestine and unregulated nature. Many ecologists have long suspected this, but new studies help to reveal why: while some species are going extinct, those that tend to survive and thrive, rats and bats, for instance are more likely to host potentially dangerous pathogens that can make the jump to humans.

### 25.3.2 WHY ARE BATS SUITABLE RESERVOIRS OF CORONAVIRUSES?

Bats are members of the order *Chiroptera* and are the only mammals capable of sustained flight. There are approximately 1,230 species of bats, making these animals the second most numerous mammals after the rodents. They have adapted easily to anthropized environments such as houses, barns, cultivated fields, and orchards, where they found a suitable ecosystem to prosper. In Southeast Asia, bats are commonly consumed as food. Bats host more zoonotic viruses and more total viruses per species than rodents, and significantly higher proportion of zoonotic viruses than all other mammalian orders. These viruses include, but are not limited to, various species of the genus Lyssa virus, henipaviruses, coronaviruses (e.g., SARS-CoV, MERS-CoV, COVID-19), and filoviruses such as several species of the Ebola virus (Keesing & Ostfeld, 2021).

Evolution has shaped the metabolic properties of bats and their immune system in ways that have made them resistant to the pathogenic effects of CoVs. The CoVs themselves have also undergone evolutionary changes that have increased their pathogenic potential: one of particular importance in the present pandemic, and in previous outbreaks (SARS-CoV), has been the ability to use the ACE2 receptors of target cells to invade them. The presence of multiple animal species in the markets in SEA and South China, including bats, produces a high density of possible susceptible hosts, increasing the probability of spillovers (Platto et al., 2021).

## 25.4 CONCLUSION

The COVID-19 pandemic is a reminder of the gravity of biodiversity loss and of our unique interconnection with nature. Despite all our technological advances we are realizing that we are completely dependent on healthy and vibrant ecosystems for our basic necessities of water, food, medicines, clothes, fuel, shelter, and energy. The coronavirus pandemic has brought these issues to the forefront of the policymaking world over. It would make much more sense to keep biodiversity, which seems to have been shunted aside by the coronavirus outbreak, in the front, in addition to various measures being undertaken to fight the pandemic.

Efforts to reverse biodiversity loss, though sometimes successful, are simply insufficient. There is an urgent need to accelerate the global response to stop upsetting the delicate balance of nature. This natural capital (forests) is key for a green recovery after the pandemic. In response to the COVID-19 crisis, UNEP is stepping up its work on mapping zoonotic threats and protecting the environment to reduce the risk of future pandemics. From reducing the risks of disasters and emerging zoonotic diseases to providing ecosystem services that are critical for the livelihoods of millions of people, biodiversity will continue to be one of the most important assets in the region to recover sustainably. The way forward for biodiversity conservation should be four-pronged, with the involvement of policy, industry, conservationists, and the general public (www.pnas.org). Additionally, host and vector management is a viable option. Other crucial measures include the restriction and control of wildlife trade, while considering the needs of indigenous peoples and local communities. Recent zoonotic pandemics have originated both from wildlife and livestock. Regulation and sanitary standards for both wild and domestic animals, and meat sold in markets, need to be strengthened. Each case requires an assessment of the best way to reduce risk while considering implications for other ecosystem functions or services.

## REFERENCES

Can, O.E., DCruze, N., and D.W. Macdonald. 2019. Dealing in deadly pathogens: Taking stock of the legal trade in live wildlife and potential risks to human health. *Global Ecology and Conservation*, 17: e00515.

Daszak, P. 2012. Anatomy of a pandemic. *Lancet,* 380 (9857): 1883–1884.

Diaz, S., Settele, J., and E.S. Brondizio, et al. 2019. *Summary for policymakers of the global assessment report on biodiversity and ecosystem services of the Intergovernmental Science-Policy Platform on Biodiversity and Ecosystem Services.* IPBES secretariat, Bonn, Germany.

Eldredge, N. 2002. *Life in the balance: Humanity and the biodiversity crisis.* Princeton, NJ: Princeton University Press.

Gibb, R., Redding, D.W., Chin, K.Q. et al. 2020. Zoonotic host diversity increases in human-dominated ecosystems. *Nature*, 584: 398–402.

Johnson, P.T.J., Ostfeld, R.S., Keesing, F. 2015. Frontiers in research on biodiversity and disease. *Ecology Letters*,18: 1119–1133.

Jones, B.A., Grace, D., Kock, R., et al. 2013. Zoonosis emergence linked to agricultural intensification and environmental change. *PNAS*, 110 (21): 8399–8404.

Keesing, F., and R. S. Ostfeld. 2021. Impacts of biodiversity and biodiversity loss on zoonotic diseases. *PNAS*, 118 (17): e2023540118.

Kilpatrick, A.M., and S.E. Randolph. 2012. Drivers, dynamics, and control of emerging vector-borne zoonotic diseases. *Lancet.*, 380(9857): 1946–1955.

Kuiken, T., Leighton, F.A., Fouchier, LeDuc, R,A.M., et al. 2005. Pathogen surveillance in animals. *Science* 309 (5741): 1680–1681.

Loh, E., Zambrana-Torrelio, C., Olival, K.J., et al. 2015. Targeting transmission pathways for emerging zoonotic disease surveillance and control. *Vector-borne and Zoonotic Diseases*, 15/7: 432–437.

Ostfeld, R.S. 2009. Biodiversity loss and the rise of zoonotic pathogens. *Clinical Microbiology and Infection* 15 (Suppl. 1): 40–43.

Platto, S., Zhou, J., Wang, Y., Wang, H., and E. Carafoli. 2021. Biodiversity loss and COVID-19 pandemic: The role of bats in the origin and the spreading of the disease. *Biochemical and Biophysical Research Communications*, 538: 2–13.

PNAs. www.pnas.org/cgi/doi/10.1073/pnas.2021460117

Robinson, M.M. and X. Zhang. The world medicines situation (WHO, 2011).

Taylor, L., Latham, S., and M. Woolhouse. 2001. Risk factors for human disease emergence. *Philosophical Transactions of the Royal Society of London. Series B: Biological Sciences*, 356/1411: 983–989.

Wang, L.F., and G. Crameri. 2014. Emerging zoonotic viral diseases. *Revue scientifique et technique/Office international des épizooties* 33: 569e581.

WHO. https://covid19.who.int/

WHO. 2015. https://www.who.int/news-room/fact-sheets/detail/biodiversity-and-health

WHO. 2019. www.cbd.int/doc/press/2019/pr-2019-05-22-idb-en.pdf

WWF. 2020. Living Planet Report 2020 – Bending the curve of biodiversity loss. WWF. www.cbd.int/doc/press/2019/pr-2019-05-22-idb-en.pdf

# Index

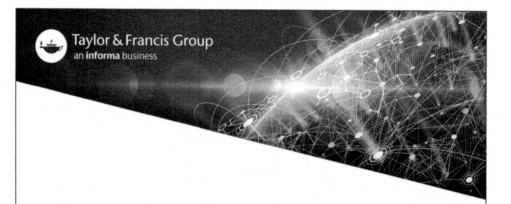